BLOOD, SWEAT, AND HIGH HEELS

BLOOD, SWEAT, AND HIGH HEELS

A MEMOIR

CHERYL WAITERS, WITH
DARNELLA FORD

iUniverse, Inc.
Bloomington

BLOOD, SWEAT, AND HIGH HEELS

iUniverse books may be ordered through booksellers or by contacting:

iUniverse
1663 Liberty Drive
Bloomington, IN 47403
www.iuniverse.com
1-800-Authors (1-800-288-4677)

ISBN: 978-1-4620-5496-1 (sc)
ISBN: 978-1-4620-5495-4 (hc)
ISBN: 978-1-4620-5494-7 (ebk)

Library of Congress Control Number: 2011916643

Printed in the United States of America

iUniverse rev. date: 10/14/2011

Exemplified by the power of the human spirit, life in the face of death and the courage to Challenge a generation to release the shackles of ignorance surrounding women and gender roles

All of this and more is lyrically conveyed in Cheryl Waiters' autobiographical novel, **BLOOD SWEAT AND HIGH HEELS. Featuring award winning novelist DARNELLA FORD.**

Profiled on **ABC's GOOD MORNING AMERICA**, Cheryl Waiters holds the noble distinction as the country's first African American woman to rise to the height of fame in her twenty-two year career in a male dominated field CONSTRUCTION WORK. With the winning combination of North Country meets Erin Brokovich, Waiters escorts the reader through a private tour of hell as she blows open the doors for an unauthorized peek inside the world of Mafia-controlled cities, labor unions, and life and death situations on job sites where women are anything but welcome.

Haunting and intensely profound, Waiters' birth and formative years are eloquently paired with historical movements that profoundly changed the world —from J.F.K to Martin Luther King, the rise of the Black Panther Movement, women's liberation and hippies toting "free love and peace, Waiters exhausts the human imagination in eye-opening expositions on American history and how they shaped and molded her to build the New American City.

BLOOD SWEAT AND HIGH HEELS delivers a message of self-empowerment for women of all nationalities and demonstrates unyielding courage to transcend the impossible and the unthinkable. The timeless genius of this story has not only captured an essential slice of history—**it has defined it.**

Given such an achievement of literary brilliance— it is destined to become an American classic.

Cheryl Waiters is currently seeking literary representation and would be happy to forward sample chapters of this riveting tale upon request.

CHER WAITERS

A third generation electrician successfully over came racial and gender bias to become the first African American female to gain historic and international recognition working in a non-traditional work environment for females and minorities, CONTRUCTION WORK.

Cher Waiters historic debut occurred while building Jacobs Field, The Cleveland Indians Baseball Stadium and the adjacent Gund Arena, currently known as the "Q" the home of Lebron James and the Cleveland Cavaliers. Cher was one of 144 women to work on a project of this size, where very few women enter the field of construction. Cher appeared on "Good Morning America" whose Joan Lunden was on a special assignment to interview women working in non-traditional work environments.

During a trip to Europe, Cher learned that the Italians worshipped a Black Madonna, remembering her fore-fathers across the Mediterranean Sea, who built the Egyptian Pyramids. Cher returned home empowered to continue to build the City of Cleveland, no matter what obstacles came her way. She never gave up. Her projects include the Good Samaritan Home, the Key Bank Tower, The Marriott Hotel in downtown Cleveland, The Great Lakes Science Center, the Cleveland Browns Stadium and many more.

Earlier in her life, Cher Waiters was told by her uncle that women did not do this kind of work. She was encouraged to go to school, get a degree, and do" women's work?" Being born into a family of construction workers and her love of math and science lead her to pursue a career in Mechanical Engineering.

During the past 25 years, the fire of commitment under President Carter's 1978 goal to hire women as 6.9% of the construction workforce, that is, seven (7) women for every 100 men, on federally funded construction projects has burned out. And now Cheryl Waiters book and movie will serve as a wake-up call that women still number less than 3% of this industry, in a warm, witty and sometimes funny narrative, **Blood Sweat and High Heels: The Journey of Cheryl Waiters, Electrician**, Cher shares her struggles and triumphs of how she fought gender based prejudice, overcame the obstacles put in her way by resentful white and black male co-workers, clung to her dignity and achieved success in this man's occupation.

Most people have to write a book before they get publicity about their experiences. Not Cher, Several years ago she was interviewed on popular television programs "Good Morning America" by Spencer Christian and "Today in Cleveland" by Del Donahue and Tom Haley. Viewers were delighted with this slender, poised black woman wearing a hard hat and a huge grin of pride about her accomplishments as she talked about being the only female journeyman electrician of her sex among 2000 male construction workers who built Jacobs Field, the Cleveland Indian's Stadium Cheryl received calls from people all over the world and was often stopped on the street and asked how she could show and help others to journeyman status and a position in construction work. This little woman with a lot of brains and courage persevered over the nay sayers to become an electrician and now earns a good living.

What one woman can do other women can do. Cher's story is an inspiration to women of all races and skills and tells what they can expect on the job and shows how to survive among macho men. Being an electrician, carpenter, bricklayer or a welder is often grueling work, but it can be very satisfying emotionally and financially if a woman knows the rules of the game, what to watch out for and how to handle the pitfalls.

Cher's story will tell all that and the need to retain the many opportunities made possible in this $400 Billion industry through federal and state programs for the so-called weaker sex to get into the construction industry. For far too long women have silently accepted that we are physically, mentally, emotionally inferior and lack the strength necessary to work side-by side with men to build our own homes and workplaces.

Cher's book explodes these myths that bad attitudes and conditioning have created. The truth is that women can strengthen their bodies and where brute strength is required, mechanical devices can be used to benefit all concerned. As well as new regulations by NIOSH (National Institute for Occupational Safety) regulations that limit every worker to lifting 50 pounds at one time. Some women regularly carry children weighing that much.

Men just do not want women in their domain even when they know they are capable, able and with proper training women can do their jobs and can do them well.

For those who seek a non-traditional career and more money **Blood Sweat and High Heels** can lead the many female heads of households who are mired in poverty down a new path to success.

Blood Sweat and High Heels is unique because it is a personal story from a single point of view and an expose' of what life is really like 24-7 for a woman who struggles on a construction site for dignity, respect and equality in treatment and pay.

"Everybody has a life . . . but the true gift lies in the ability to express that "life force" in a way that is thought provoking, entertaining, inspiring and educational to anyone who might see that life. This life then becomes more—it becomes art."

Cheryl Waiters

Preface

GOOD MORNING AMERICA INTERVIEW
Featuring Cheryl Waiters
April 4, 1994

JOAN LUNDEN
It is Cleveland's new field of dreams . . . Jacob's Field.
Spanking brand new
Ready for opening day
Men and women have built it . . . now they wait for people to come.
Good Morning America!
I'm Joan Lunden

CHARLES GIBSON
I'm Charles Gibson . . . good to have you with us.
It is Monday, April 4th
This is baseball's opening day,
And a brand new ballpark in Cleveland
Ready for the first pitch of opening day,
With the Indians and the Mariners.
Cleveland has a new home, a new division,
A new era for baseball.
Also new, women in the construction gang
That built that ball park.
A little later in this hour Spencer Christian
With some of the women who labored to build
Cleveland's new field of dreams.

[Later in the hour]

CHER
Hi! I'm Cher Waiters.

MONICA
I'm Monica Jordan and we built Gateway.

CHER
Isn't it beautiful?

CHER AND MONICA
Good morning America!

CHARLES GIBSON

That's amazing that just two women built
That whole thing and got it ready for opening day today . . .
Gateway is the overall complex in downtown Cleveland,
And the critical part of it is Jacob's Field where the Indians
Will open this new baseball season, as we have been
Talking about this morning,
There is a lot new about the 1994 baseball season
That gets underway in earnest today, but as columnist
Tom Bosweld pointed out earlier—
A lot of what's new is really old, such as the modern stadiums in Texas and Cleveland opening this season . . .
But stadiums that really pay tribute to the past and really a good example is Cleveland's new Jacob's Field . . . with its natural grass and freestanding scoreboard makes you feel like you're part of the game that really really is very old, and of course baseball is, but the construction technology—while it's new
So is the makeup of the workforce that built the
Stadium new.
Women now dawn the hard hats along with the men.
In a few moments Spencer Christian from Cleveland
With the new breed of workers who built the new, old ball yard.

SPENCER CHRISTIAN
And we'll be back from Cleveland's Jacob's Field
In just a moment with two of the female construction
Workers who helped build it when Good Morning
America continues.

FADE OUT.

Chapter 1

"Being born is like coming into the middle of a movie. You have to find out what happened before you arrived and catch up to where you are now."

Cheryl Waiters

This journey begins before time—at least before *my* time. It would be impossible for me to dismiss the impact of all that came before me. Undeniably, the current events of yesterday left a great impression in the form of an "invisible dent" upon my human skin. Disregard the fact that I have not yet been born—it is only a technicality. Sooner than you can recite your first name backwards, I *will* be here. And undoubtedly, I shall feel this "dent" upon my arrival.

But not yet.

I'm not yet here—

Remember?

It was the late 30's—an era memorialized by The Great Depression. During this time, no man was beyond the reach of a twisted fate, and even the fortunes of the wealthiest citizens ran bone dry all the way down to the spit in their mouth. Deep pockets flipped outside in and became incredibly shallow and as for shallow pockets—well, they just blew away in the wind.

The face of strangers and friends reflected the hollow echo of emptiness—empty hearts and empty bellies. Entire families waited in line all day to hold a crust of bread in one hand. There wasn't enough food to fill up two hands.

It was a time when it would seem that Mother Earth had taken great offense toward humanity. The sky stopped crying and the rain stopped falling. In exchange, the people began to cry because everything started to die. Farmer's crops dried up in a bitter drought, thus giving birth to a time where the land delivered nothing but anguish and more anguish. This period, known as The Dust Bowl or the Dirty Thirties, began in 1930 and lasted till 1936. That's not to say that the Dirty Thirties were all bad. The era did deliver some redeemable achievements to press against our beating hearts amidst the pain.

The decade of the 30's saw the construction of the world's tallest building, the Empire State Building in New York City. Picasso's "Guernica" was the talk of the art world and Amelia Earhart disappeared into thin air. Hemingway and Steinbeck were the celebrated authors of the time and *Gone with the Wind* was released and took everybody's breath away. The 30's reigned supreme when the smallest planet in our Galaxy, Pluto, was discovered. The miracle drug, Penicillin, was introduced in clinical drug testing and Diabetics found a new reprieve with the modern day miracle of Insulin.

There was yet another significant event which occurred during the era of the '30's—and though it did not make the history books; it became a pivotal catalyst to my arrival here on planet Earth—the birth of my father, Charles Waiters.

Born in 1937 in Columbia, South Carolina, Charles was a Depression Era baby. Therefore, it goes without saying that he was stamped with a hunger for wealth. Branded by this insatiable passion for the finer things in life—luxury was encoded in my father's DNA. Almost traceable, it was part of his genetic code and this wild ambition would rule him for all the days of his life.

Charles Waiters longed for the best of everything, and he would not stop short of having it—*all.* The words "to settle" never made it to Charles' vocabulary, therefore, he did not understand the meaning of the words "to have less than one deserved." He was quick-witted and well-armed with just enough of everything he needed to acquire everything he ever wanted.

Yes, that was Charles Waiters.

The son of a well-to-do Mason and offspring of a family whose business was construction—Charles had two brothers, Willie and Alfred. Now most of the time Charles got along just fine, however, he never managed to get too far along without running up against somebody's comparison of him against his brothers.

Willie was the smart one.

Alfred was the pretty one.

Charles was neither.

There were also two additional brothers and a sister, but they seemed like "ghost siblings" whose impact upon the life of Charles was minimal.

He wasn't a bad looking guy but he was a long way from being pretty. And by no means was he a dumb man—but he was a "double-wide hop-scotch-jump" away from the category of genius. And it was from this place that Charles had to learn to make his way in the world. In doing so, he fell in love with gangsters. He had a thing for Frank Sinatra and an affinity for Hollywood's portrayal of the mob. Instinctively, Charles knew that if he couldn't be pretty or smart, that he could be a self-proclaimed "elegant gangster" and ultimately, he could get paid. Like I said before—he was tattooed with the thirst for wealth.

The Waiters migrated from the South to the North like many Southern families—seeking their own middle-class version of Utopia, or at minimum a less oppressive society from the white man's world. The land of the North seemed to offer a seductive whisper to those who had the ear to hear the melody of freedom. Grandfather Waiters heard the call in 1948 and relocated the family from South Carolina to the promise of a new paradise: Cleveland, Ohio.

Who knew?

The Waiters took up residency along Cleveland's Gold Coast and lived on the "right side" of the tracks. A quiet, conservative and respectable family—they lived under the intention of honor, however, the decade of the 40's gave birth to a time when everyone encountered their own set of challenges—good intentions or not.

It was an era of war, killing and bloodshed.

It was pure insanity that history buffs politely label as "global conflict."

World War II.

It seemed to be a time when the sadness of the 30's was replaced by the tears of the 40's—though not everyone was crying. In 1947, we saw Jackie Robinson, the first African American baseball player enter the Major Leagues. And then there were less notable accomplishments, though they managed to change the face of America in their own way.

Sweaters became a popular fashion statement, though there were vicious debates on the "moral" aspects of a woman wearing such a tight-fitting garment. The Roller Derby was in full swing as a professional sport and boxing was gaining great notoriety with people from all walks of life. And yet again, there was another event which quietly escaped history's notice—the birth of my mother, Mary Dunham.

She was born in 1941 in Columbus, Georgia.

A beautiful high-yellow woman with a flair for "sexy," Mary also migrated from the South to Cleveland, Ohio with her family when she was a young girl. However, unlike the Waiters—my mother would live on the "wrong side" of the tracks. She was the offspring of a humble people with little financial means, but they did have a whole lot of Spirit—in the form of religion *and* booze. It would be a stretch to call them conservative, however, Mary and her family did believe in the 'Good Book' and the 'Good Lord' whenever it was fashionable to do so—which was mostly on Sundays. But if you were to catch them on any *other* day of the week, the whole damn thing was subject to collapse under the weight of a single question:

What in the hell?

Mary's family was loud and they loved to have a good time.

Discipline was not a practiced lifestyle and short of everything—pretty much anything else could go over without a challenge. And as my father grew into early manhood and my mother blossomed into the peak of her teenage years, their paths would cross.

Charles was from the right side, and Mary was from the wrong side; however, they met on neutral ground—*The Red Carpet*—a hot spot in the Cleveland nightclub scene.

Mary, who was 15 at the time, had gone to the club that night with her friend, Louise, who was also 15. The two girls seemed to have a certain destiny with Charles, a savvy nineteen-year-old, and his brother Willie, who was the eldest at twenty-two.

According to Willie's version of the story, Mary and Louise were raging with hormones and seemed a bit "fast" for the two young men. Charles and Willie admitted to being hesitant to have any dealings with the young women, however, "boys will be boys" and the biological call of a man's urge to mate—at times make a difficult case for the argument of good old fashion common sense.

Six Months Later

The year was 1957 and a new decade had been born. Two pregnant teenagers, Mary and Louise, stood in the middle of a chapel under the harsh eyes of displeased parents in a quietly celebrated shotgun wedding. It was a time of reform and great change, and not just for Mary and Charles or Louise and Willie, but for everyone.

The stirring motion of unrest had begun to swirl all around America in the 50's and as we edged toward the 60's, we tipped the scales toward revolution. Our parents didn't start the revolution—they simply carried out what had already begun by those who had come before them. It was those of our grandparent's era who ignited the great winds of change. Spirits had long grown restless, and change was *everyone's* destiny as Sam Cooke predicted in his 1963 hit song, *"Change is Gonna Come."*

Change was going to come all right and by the 60's that change was well underway. But prior to the dawn of a new day, Charles and Mary Waiters were caught up in their own private revolution—becoming teenage parents. A child in her own right, my mother was content to trade her prom gown for a maternity dress and the single life for diapers and baby formula.

Saddle up Charles, we're signing on to play this game for life!

Just following their wedding ceremony two suspicious men appeared at the church.

"Who are those men?" my mother inquired.

"Nothing to worry about," Charles assured her.

"Man . . . don't you have no respect?" his brother, Willie pressed.

"I gotta do what I gotta do!" Charles insisted.

Though he worked as an auto mechanic with decent wages, it would never be good enough for a pure bred like Charles Waiters. After all, he did have royal blood coursing through his veins and it only responded to the "finer things" in life, remember?

Following their wedding ceremony, Charles and Mary returned to his mother's house. Short on money and even shorter on dreams, a small bedroom would be their honeymoon suite. But they didn't spend their first married night together in bliss. Instead, Charles excused himself because his "other gig" was calling for his undivided attention. He was taking it to the streets—for every "elegant gangster" knows there's no such thing as an off day—even on your wedding night.

Long and short of it—this was my parent's story.

This is how they began.

Not how they ended.

Their tale is woven with a trail of blood, sweat and tears.

Almost *everyone's* is.

Everybody has their own story to tell, including America, who was birthing a tale of its own prior to my arrival—as these were the highlights in American History during the 50's:

Eisenhower held office as the President of the United States.

The Immigration and Naturalization Acts were passed removing racial and ethnic barriers on becoming a United States citizen.

Racial segregation was ruled unconstitutional in the U.S. Supreme Court.

Rosa Parks refused to give up her seat on a public bus in Montgomery, Alabama.

Alaska and Hawaii became the 49th and 50th states.

Jerry Lee Lewis said it best in 1957 when his words echoed throughout America, "Whole Lotta Shakin' Goin On."

The 50's were a profound era no doubt—but make no mistake when rewinding the history books—it was the *music* that defined the era. Transcendental by nature, music changed the face of racism. Powerful musicians and innovators such as Chuck Berry brought the borders of segregation all the way down to its knees. The ignorance of man, whether it was self-taught or learned from another, could not withstand the beat.

Body and soul, black and white alike—we all surrendered to it.

We gave in to the sound of music, and in doing so—we overlooked the color of the man who played it. Little white girls went crazy for darked-skinned boys toting straight perms, guitars and fancy footwork.

It was nothing short of miraculous.

Music bridged worlds and in the process built new ones.

It transmigrated the soul of culture itself.

It changed the face of our planet *forever.*

And somewhere along the way—long after the Emancipation Proclamation and during the raging storm of The Civil Rights Movement—at the end of one rainbow and the beginning of another, I entered the movie of life.

May 25, 1958.

My name is Cheryl Denise Waiters.

I am Black and female.

Big disappointment.

Everybody was praying for a boy.

There will be hell to pay for not having a penis in this lifetime.

But I am fresh blood.

I am also the dream of America—temporarily. For one day, I shall become the nightmare they all wish to forget about.

This is my tale.

These pages contain my blood.

My sweat.

And my tears.

My history as *her* story.

My story *and* your story.

This is our story.

Good Morning America.

Chapter 2

"Women are the co-heirs of the Universe."

Cheryl Waiters

It seemed at birth that I was destined to learn the answer to a question that I never wanted to ask—but Mother Nature invested a sincere interest in the debate of gender difference, perhaps for no other reason than for the sport of it.

". . . what's the difference between a boy and a girl?"

"I beg your pardon," I kindly replied.

"Boy versus girl," Mother Nature insisted.

In my humble opinion—the answer was much more than anatomy.

It was not his penis.

Nor was it my vagina.

No, it was much more than just biological.

. . . what's the difference between a boy and a girl?

The answer—*everything.*

That's the difference.

Women are the co-heirs of the Universe and we deserve our rightful place. However, history long in the making transcends time, space and race to quantify the difference between a boy and a girl—which far surpasses the notion that one shall play with fire trucks and the other shall play with Barbie dolls. Male children are not only favored, but highly preferred in many societies. A peek behind closed curtains reveals the tale of the People's Republic of China where the government not only introduced, but enforced with penalty, a policy which allows married couples to have only one child.

Imagine that.

And it still holds to date.

As a result of such an order to control their birth population, people became desperate and obsessed with male babies. If they only had one shot to reproduce—they didn't want to waste it on a girl. There was a mass genocide of infant girls and late-term selective sex abortions performed on female fetuses, which ultimately created a gender imbalance in many parts of the world.

In Kabul, Afghanistan in a custom that reaches back through centuries and is still practiced today—desperate families began dressing their girls in disguise as boys. They are known as *"bacha posh"* which means "dressed up as a boy" in their native tongue of Persian Dari.

In many cultures, boys are favored because they carry the family name—while a woman who produces "girl after girl" is often viewed with a mixture of contempt as well as pity. Modern day research suggests that almost 100 million women are missing from this planet as a result of selective sex abortions and female infanticide.

With all of that being said—I am most pleased that I made it.

In 1958 in Cleveland, Ohio where I was born—such extremes were not a part of America's cultural norm, however, they *were* a part of the culture. So at the end of the day, my parents weren't much different in *attitude* than those of the People's Republic of China or the city of Kabul.

They, too, wanted a boy.

Desperately.

But they made due with the disappointment of having a girl.

"We'll try again for a boy," my father, Charles, reassured my mother, Mary.

There was no time to mourn what nature had delivered in the form of female genitalia. My parents had to get on with the business of living, because responsibilities had to be met one way or another.

Girl.

Boy.

Or other . . . the rent still had to be paid and a way still had to be made, but my father was not a fan of structured employment. That's my way of saying he bounced around a lot and words like "consistency" and "longevity on the job" were as foreign to him as the Cuban Missile Crisis of the early 60's.

If Frank Sinatra didn't have a 9 to 5, neither should Charles Waiters—in his own mind.

"I am an enterprising young man!" he boasted.

Perhaps my mother bought in to the spirit of his entrepreneurial ways until the doorbell rang one night at the home of my grandmother—where my parents were still living after my birth.

"Who's there?" my grandmother inquired.

"FBI!"

Well, when you hear those three little letters strung together—they make for a mighty LOUD sentence, even if it's said quietly.

"FBI?"

Upon entry, the two official-looking men flashed badges and unfriendly smiles—if you dared to call it that. After that, they got straight to the point. "We are here on official business surrounding Charles Waiters and we do have a search warrant!"

"Sweet Mother Mary of Jesus!" my grandmother exploded into a panic, but all that she could do was step aside and not interfere with what was already taking place. A thorough search only turned up a bit of marijuana in Charles' coat pocket, but the FBI was investigating him on more serious charges of narcotics distribution. Needless to say, this did not sit well with Charles' mother or with my mother.

"I want you to get a job, Charles Waiters!" demanded my mother.

"I got a job!"

"Dope dealing ain't a reliable source of income," she insisted.

"I'm doing what I gotta do for this family!"

"What you gotta do is get you a real job! You got responsibilities now!"

In one ear and out the other—that's where Mary's words went. Charles Waiters would bow to no man or woman, and whether it was his wife or his mother, it wouldn't make a single bit of difference. At the end of the day, all words were empty words and they fell upon hollow ground. But Charles didn't care—he'd step right over the top of every noun and verb even from the most compelling of lectures.

Scholarly advice didn't move him.

None of it did.

He would not allow the world to give him *its* terms—he was hell bent on giving the world his own terms. After all, wasn't it Frank Sinatra who popularized the Paul Anka song, *"My Way?"*

And Frank was Charles' hero, remember.

Six months later Mary was pregnant and "responsibility" was growing by the minute *and* by the month depending upon who was taking the measurement. And for all intense purposes, there was a growing measure of responsibility that both of my parents were stepping into. Therefore, real jobs needed to be found and all lollygagging had to cease immediately. Eventually, Mary took up a trade and became a Candy Striper, a hospital volunteer. Charles continued to do "odd jobs" here and there—but again, consistency would remain the biggest threat to the peace of the household.

At the age of 12 months, I became a "big" sister.

Alas, Mary gave birth to a baby boy.

Charles had struck either gold or oil—it didn't matter which it was because both were more valuable than a female child. My parents were thrilled to have a son. It was as if the Gods had shown them special favor. Naturally, he was named after my father—Charles Waiters Jr. He was just what the world needed—another elegant gangster, one who communicated by way of high pitch squeals and low pitched coo's. My little brother was an interesting character study indeed. After Charles Jr. was born there were changes to be made—slight adjustments or major overhauls depending upon how you looked at it.

We needed more of everything.

Space.

Time.

And money.

Ultimately, Charles and Mary moved out of his mother's house and got their own place. I have very limited memories of us all being together as a family unit, but one memory I do recall quite vividly—my mother came home one evening from work. She was wearing her uniform and a big ugly frown which was stretched across her beautiful face from ear to ear.

Uh oh.

Trouble in paradise.

That's what that look said to me, even as a child.

"Charles Waiters!" she declared in a booming voice. "You were supposed to pick me up from work! What happened???"

I don't remember what my father said or maybe he said something as simple as, "I didn't remember," but that blank stare across the top of his eyelids was definitely no reassurance that he had a good answer as to why my mother had to walk all the way across town in the cold weather. It was one of those unpleasant kind-of-family situations that seemed to advertise without compassion:

Damned if I do.

Damned if I don't.

Next thing I knew—Charles Waiters was moving out and he and my mother were on the out—out—outs! I was three years old at the time. At the age of three—no one seeks your counsel or contemplates giving you two good seconds of an explanation as to what the hell is really going on.

There is no putting it in "little people's" terms.

You go to bed one night and you wake up the next morning and the world is different.

Big people don't think little people know the difference, but I beg to differ—little people *always* know the difference. In fact, they surpass big people in intelligence most of the time, simply because they don't assume that they already know everything; and in that assumption, they leave the door wide open for further learning. Little people may not always express their feelings—but that doesn't mean they don't feel them.

So like I said, I awakened to a new world.

It was irrevocably silent.

Still.

Almost motionless.

The adults in my world continued to mimic the same steps of yesterday. They moved in the same manner and even did the customary dance they had always done—but the music had changed and the beat was all together different.

They knew it.

I knew it.

My little brother even knew it but we pretended not to notice. Sometimes, pretending was what we did best. It was as if we had previously moved to the beat of rock 'n roll and then awakened one morning to a room filled with the twang sounds of Country music.

Who wouldn't notice that?

And even more peculiar—everyone was still grooving to a rock 'n roll beat.

They're all nuts!

At least I declared them to be so.

The adults had gone mad and no one noticed except the children.

When my father moved out the hole widened in my heart. I missed the Son of a Bitch something fierce. Perhaps I have inserted the word *Son of a Bitch* here as a term of affection between a father and his daughter—or maybe not.

Maybe he was just a Son of a Bitch for no other reason than that's just what he was. Don't get me wrong, my father came to see me and Chuckie from time to time. His guest appearances were anything but consistent but they were always grandiose. My father was larger than life so when he came around it was as if he had never left because when he was in the room—he took up all the space in it. I wanted to breathe him in. He seemed to occupy every corner of my existence and I wanted to be everywhere he was. I wanted to see everything he saw. I wanted to hear everything he had ever heard and when he schooled me on the finer things in life, I always paid attention.

"Money's my best friend," he always boasted, "my very best friend."

From an early age, I could see that my father and money had a sweet kind-of-relationship, and I wanted in on *that* kind of relationship. My father bled the love of money in through my veins and it became a part of my blood.

I loved it.

I wanted it.

Ached and longed for it.

My father was an elegant gangster and I was his daughter. My eyes couldn't help but catch the glitter of his diamonds as he had a flare for the exquisite details of life which he kindly expressed through finely pressed suits and a big pretty smile.

"When you coming back? When you coming, daddy?" I whined and begged for more of his company.

"Real soon baby girl . . . real soon," he always promised.

Some of the promises he made good on, but most of them he did not. I never got used to the disappointment of his broken words. I had always hoped that somehow the words would fall out of his mouth and fix themselves and just "be right."

Do the right thing, Charles Waiters.

Do the damn right thing.

I couldn't get enough of my father.

I had an outrageous, over-the-top, unconditional, insatiable and unquenchable love for the man known as Charles Waiters. He didn't deserve it—but he got it anyway, because I am a kid. I was *his* kid.

And when he did make an appearance—he made it in a red convertible. It was a real pretty car with a whole lot of shine. Ironically, it was just like my father. Whenever he came around he was toting good cheer and boasting of American prosperity—all the while teaching lessons on life.

Grab the tiger by the tail . . . hold on and don't let go.

Be number one in all that you do.

Save your money.

If you want a Cadillac, don't talk to the man with the 1957 Chevy.

I bought his words of wisdom hook, line and sinker.

Daddy had advice for days, but it was not until later in life that I would discover most his advice was flawed to its utter core.

My mother, on the other hand, was *not* a fan of Charles Waiters and after he left—bitter sang its own tune for days. Proud and intentional, Mary Waiters never asked Charles Waiters for a damn thing. In that, she took what little he gave—if anything at all and cheated us right out of our mother loving inheritance.

Chapter 3

"When the power of love overcomes the love of power, the world will know peace."

Jimi Hendrix

The script flipped when Charles Waiters moved on about his business and further away from the interior of our lives. Within the quarters of our humble living environment—the rumblings of Revolution were brewing. As a child, I could feel the disturbance rising in the air but I just didn't know what to call it—eventually I would learn.

It began as "atmospheric pressure" in our home—when the adults would lose their temper over things that had gone on for too long. In that space, a lot of words were said but not much was communicated. There seemed to be a gap between listening and hearing, and an even wider gap between hearing and understanding. The blind was surely leading the blind which meant everybody was bumping into walls. My parents were probably no different than a lot of the parents back in the day. They were good at sweeping things under the rug—but only until their passive approach to life accumulated enough debris to cause a wild dust storm.

Run for cover . . .

Somebody's about to blow their stack.

Damage control was a necessity especially when forced words came out easily wounding those who happened to be in the way. Relationships always ended with a countless number of casualties despite everybody's "good intentions" on not getting hurt or hurting anybody; and even the most functional of families were dysfunctional at best.

Welcome to my world.

My family was no exception to the rule of genetic dysfunction.

Ironically, the exterior world mimicked our internal environment. It was the early 60's when my father moved out of our home. He left just in the nick of time for all hell to break loose in America. The world was bent on Revolution. The 60's lit the world on fire, and I was a child of the children who were steering a course already set in motion toward change.

At the age of 43, America's youngest president, John F. Kennedy, was sworn into office in January of 1961. Martin Luther King Jr. captured the attention of the nation when he marched through the streets of Washington D.C. in 1963. And perhaps we all believed that we had overcome till Kennedy was killed in November of 1963 and then we all felt like we went backwards again. The Beatles touched down in America in 1964 and changed the world forever with their music. During this time Motown founder, Berry Gordy, was also rewriting history and breaking through profound color barriers with the discovery of musical influences who not only helped to shape the world but change it significantly with acts such as The Supremes, The Temptations, The Four Tops, The Commodores and The Jackson 5.

President Johnson declared an all out "war on poverty" in '64 and the Watts Riots hit in 1965 and nearly burned the City of Angels down to the ground. And somewhere along the way—there was Civil Rights, Black Power and the Black Panther Party which was founded in Oakland in 1966. It seemed that the world would wither under the weight of both Martin Luther King and Bobby Kennedy's death, as both were assassinated within three months of each other in 1968. America cried out—both black and white alike. As a nation we didn't know where we were headed. Ironically, we ended up on the moon in 1969 when Neil Armstrong took "one small step for man and one giant leap for mankind."

Hell of a time to be alive. It was a time of widespread segregation and people didn't hide their hatred of other races. Opinions ran rampant and no apologies were made. Coloreds and Whites stood in two separate worlds while standing on the same ground. Aggression through a force of creation—the 60's represented times of change en mass. The Vietnam War was in full swing, but the pendulum swings both ways because it was also about the love with the emergence of a subculture of Hipsters *(i.e., Hippies)* who ran counterclockwise to society's code of standards at the time.

Peace . . .

Love . . .

And Woodstock in 1969.

Hippies were everywhere toting flower pants, long hair and hearts which bled of compassion. They seemed more at ease than the rest of society—could have been the massive amounts of ingested pot.

Who knows?

I'm not here to judge.

They made their statements through the exploration of a lifestyle which expressed itself through clothing, art, music, culture and did I mention pot? With the introduction of the birth control pill in the early 60's, women became sexually empowered and liberated to engage in more risky sexual behavior. I would imagine this shook things up and got them started, and either way you look at it—the world was turning upside down or right side up—depending on whether you were being born or dying at the time.

This was the era that breathed me.

Shaped me.

And molded me.

I was dented with the impressions of the 60's and these dents blew holes and doorways into the future.

My future.

Spoon fed images of My Three Sons, Star Trek, The Twilight Zone, Dick Van Dyke, The Little Rascals, The Beverly Hillbillies, Gomer Pyle and Andy Griffith—I was looking for my own version of Mayberry. The living was easy and I would have been a fool *not* to look for it.

Everybody smiled, waived good morning to their neighbors and wore nerdy haircuts. The surplus of life's wealth were sprinkled everywhere—at least on television it was.

I wanted to live there.

And where was there?

But of course . . . where all of the white people lived. I wanted to shop in their grocery markets and ride in their Chevrolets. It was the sweet, hushed sounds echoing the gentle lullabies of White America; and just like Michael Jackson, I, too, wanted to be white.

Born in the same year of my birth, I understood Michael.

This is what we saw.

This is what we were told.

And this is what we were sold.

As I said before, I bought it hook, line and sinker. But in the words of John Lennon, "I'm Not the Only One . . ."

Would it be self-righteous of me to presume that you now understand where I come from? Or at least from my perspective, have a measure of compassion for where I may be headed? From hell all the way to "here," which happens to pour itself out in the heart of a city disguised as "eternal damnation," otherwise known as Cleveland, Ohio.

When my father exited stage left, my mother took center stage in assuming both the role of mother *and* father. A poorly acted role, I oftentimes forget that she was a mere teenager when she stepped in to play her part. By the time she was twenty years old, Mary Waiters was already a single mother with two children and a whole lot of growing up on her own left to do. She expressed no natural inclination towards Academia—therefore she put no real effort into education and its potential benefits. Furthermore, she put no real effort into motherhood or in pursuing Charles Waiters and force feeding him a single spoonful of "responsibility" in taking care of us.

So exactly what was Mary Waiters interested in?

Partying.

Playing.

And being anything it seemed but grown up—maybe because she grew up too fast.

It's funny how life works that way—the fast girls grow up "fast" because they want so badly to be grown, only to regress to child-like behaviors once they reach adulthood. It all happens quickly and before you, me or *they* know it—the party's over, the kids are here, diapers need to be changed and mouths need to be fed.

"I'm going out tonight," I overheard my mother say on the phone, "but we'll be at church tomorrow."

My mother was devoutly religious. She carried her Bible in one hand and the password to the Afterhours club in the other. It was an interesting blend of two entirely different worlds. Mary did what she could to take care of the basic needs of my brother and myself. The rest was a crap shoot. Don't get me wrong—we were fed, bathed, clothed and fitted with a roof over our heads. Our physical needs were met. Mary always saw to that, but there was something vacant or simply gone missing in the spiritual and the emotional. My soul needed to be nourished. It felt bone dry. I needed somebody to pour some oil over my head and fill me up with something soothing.

We are hybrid beings—part physical and part spiritual. Humans feed the physical just fine *(i.e., our clothes, our food, and material things)* but we neglect the spirit, which is the fuel for the physical. That part of us is lacking and failing. When that is not fed our entire mechanism will fail.

From my earliest of days, I felt like a poorly nourished vessel—fattened up on the hype of the 60's. I was being "etched" into life as a militant as I consumed the invisible waves of aggression which offered quite visible results of their manifestations in real life. The world was going crazy with uncivil rest at every nook, cranny and corner. Dark caverns of the human mind were being explored without adult supervision. Television was brainwashing us and everything around us seemed to reinforce the bill of goods we were being sold; therefore, we just kept on buying it without question.

"Don't trust Whitey," my uncle used to tell me when I was a kid. He was a member of the Black Panther Party. I was scared of him as a child. His militant stance, intimidating eyes and crinkled brow never seemed to make for an inviting conversation. With that being said, I was primed to be a "fighter" before I ever entered the ring. Socially, the routine was already choreographed. Society and well-intentioned family members would see to it that I grow up respectable with a balance of good manners and etiquette. A well-behaved child, I took the golden rule to heart. And I also took all of the rest of the rules with me for good measure and stuck them in my back pocket:

1. Respect your parents
2. Respect the elderly
3. Respect other people's property
4. Be considerate of others

My generation was the last to be taught such etiquette for living—and it is for that very reason that I refer to us as The Last Mohicans. We are the last of a dying breed and we bear the imprint of the revolutionary 60's on the breastplate of our souls. However, the branding was unsupervised. Ultimately, my parents and probably many other parents were too young *and* too naïve to keep a watchful eye on the indentations that were pressed down upon the souls of the children who were growing up amidst the Revolution. And really . . . how could they keep a watchful eye? Hell, they were just trying to pay the rent.

Chapter 4

"There is always one moment in childhood when the door opens and lets the future in."

Deepak Chopra

Childhood was an interesting experience. I offer extensive commentary on the reflection of my early years as there was always a "self-admitted" struggle to balance love against the anger—in which I felt generous amounts of both emotions for each parent. And once I reached adulthood, I gave my parents a "final" report card which summarized my viewpoint of their parenting skills. Interestingly, if we appraise their skills as lofty—then perhaps we are well-pleased with their efforts and applaud their contribution to our lives. If our review is an endless list of unforgiving citations for the shitty job they did as parents—well . . . I guess this is where we find the "grown up kid" sitting on an uneven sofa in some downtown shrink's office crying crocodile tears over spilled milk that's still spilling.

Welcome to my world.

Again.

Growing up the eldest and only daughter of Mary Waiters brought its own unique set of challenges. I was not an ordinary kid—at least by my standards of ordinary. I did not wait until I was an adult to begin my parent's "final" report card. Interestingly, it began at a very young age. The moment I came through the birth canal, I began my appraisal—a checklist of satisfactory versus unsatisfactory behavior which gave way to a "Pass," "Fail" or plain old

"Just Try Again." I wanted to give them both A's. I was always enthusiastic on the subject of great accomplishment. I was hell bent on it—not just for my parents but also for myself; and even as a toddler I trace the memory of profound moments which served to lay an early foundation of personal achievement. As a "tot" I vividly recall sleeping in a crib with extremely high bars surrounding it. Designed for my safety, it felt more like a prison with its caged restrictions holding me steady and in one place. I didn't like the crib. I had no intentions of being held in place—at least not for long anyways.

On this particular evening, I remember the moon shining brightly through the window illuminating everything in sight. It almost seemed as though the "inanimate" objects came to life and I stared in awe of them. In the background, I could hear my mother and my Uncle Marshall sitting at the dining room table talking and laughing.

"So what you been up to Marshall?" Mary asked.

"Working man," he replied, "always a working man."

My eyes lit up at the sound of Marshall's voice. It was a trained response because Marshall always came bearing gifts in the form of candy, and I wasn't about to miss out on my treats. So I plotted a carefully constructed plan to get out of this "contraption" by stacking up pillows and blankets on one side of the bed—building enough of a platform to hoist myself up and over the bars.

Ah ha!

Ingenious!

Or so I thought, but only till I realized that hanging over the other side and dangling off the edge of a bottomless floor was *not* what I had in mind. I can remember my little feet—which were nestled inside of the little "pajamas with the feet in them" as they hung desperately in the air.

Uh oh.

Now what?

I tried to look down to the floor to see where I had gone awry with my estimate on the distance from the crib to the floor, but it was too dark in the room for an accurate measurement. It seemed as though the black bottom of infinity would swallow me whole should I decide to let go. Therefore, I hung on for dear life.

Terrified.

What could I possibly do?

Don't be scared, I told myself.

Let go.

On one hand, I didn't believe a word I had just told myself, but on the other hand, since there was no one in my immediate future to come and "rescue me," letting go sounded like the most reasonable option I had in the moment.

Let go.

Slowly, I released the white-knuckled, death grip I had on that old school crib and soared toward the center of the earth at what felt like rocket-speed.

Crash landing!

I hit the ground hard but I got up running and was filled to the brim with ecstasy that I was still in one piece.

I did it!

I transcended my fear and defied gravity all in the same breath! Okay, maybe those weren't the exact thoughts that came to me, however, a surreal sense of accomplishment and something that felt superhuman were in my mind and I got high off of it.

Yes!

I could encounter a challenge, devise a solution and execute a plan to utter perfection. I figured this out as a toddler.

Imagine that.

I could win!

In that moment I knew I could do anything. Knowing this filled me with excitement. I wanted that feeling again—the rush of a pure sensation. It could be likened to a homerun, touchdown and a slam dunk all in the same breath. I ran straight out of that moonlight bedroom and into the kitchen where my mother nearly toppled right out of her chair when she saw my beaming face, but by then it was too late—I was already jumping up into Uncle Marshall's lap looking for my candy.

"Here there pretty girl!" roused Marshall. "Where did you come from?"

"And how did you get out of that crib?" Mary squealed with a perplexed look on her face.

I laughed out loud. *I'm smarter than you are,* I wanted to say. At the end of the day, I got my candy and that was all I really cared about.

Well not really.

I cared about others things too—like the fact that this was the first time I did something greater than me. That feeling stayed with me for years. Even as a child I reasoned that if I had done it once—I was destined to do it again.

And again.

And again.

Again would come later.

First things first.

Mary Waiters was a single woman with a dual desire: 1) she needed a man 2) she needed some money. I was never sure which one she needed more—maybe *both.*

When I was five years old Mary met a new man—a snappy dressing, chocolate brother who went by the name of Ronald. At first glance, I didn't have a problem with Ronald, but that's only because his problems couldn't be detected upon sight.

"Hey sweet thang . . ." he said extending a hand to me, "my name is Ronald. I'm your mother's friend."

"Hi," I said timidly, checking him out from a distance.

"I don't bite, pretty girl . . ." he said. "You can shake my hand."

Slowly, I accepted.

"What's your name?"

"Cheryl," I responded softly.

"She ain't shy!" my mother spouts. "She just playing shy."

"She just playing," my brother Chuckie chimed in.

"What's your name, young man?" Ronald asked, turning his attention away from me.

"Chuckie."

"Nice to meet you, Chuckie."

That's how it started . . . pleasant enough or so it would seem. Ironically, it didn't end that way—guess there's always another story that's told somewhere between the beginning and the end. It could be a dream or a nightmare depending upon who's telling the story and also who's living it. Ronald and Mary were quick in setting up shop. Hell, next thing I knew he was moving in and declaring himself, "man of the house."

Really?

Who knew?

My father had all but disappeared into thin air and mostly resembled a ghost these days. His visits were sporadic and highly unpredictable, if at all. When my father moved out—my uncle Willie (Charles' brother), stepped in to try and fill some of the space his "empty shoes" left behind in our living room and in our heart. I loved Uncle Willie. He was a practical man who had his own thriving hustle as a working electrician. He didn't work for the man. He worked for himself and I liked that. Even as a little kid I saw the value in being self-created.

Family life may have been lacking in certain areas, but once Mary's steady beau, Ronald, moved in—it seemed as though the atmosphere changed. The breathable air in house felt like it was being doled out in shorter supply.

Specializing in "criminal activities," Ronald always had a healthy stash of cash. And he didn't mind sharing so he often came bearing gifts to me and Chuckie—but I wasn't as impressed with the "glitter and gold" as I could have been—because he also came bringing in "extra" ticket items that we did not request. Ronald brought violence to our home.

"Woman! I told you what I wanted you to do!" he screamed at my mother.

"Ronald . . ." she pleaded in her defense.

She would try to reason with him—but her reasons never went far in translation. A simple exchange would escalate until it became a critical boiling point, and then the rest of the conversation was inserted between blows and jabs.

Smack.

Punch.

Shove.

Ronald didn't play fair and he hit below the belt. When he wasn't in the midst of creating chaos and drama, need I say that he was the neediest Bastard in town and all but sucked the red out of my mother's blood. Incessantly in need of something, he occupied all of my mother's time.

Attention.

And affection.

The well ran dry and there was nothing left for Chuckie and me.

I didn't care for Ronald much at all, however, I did tolerate him as best as I could.

At the age of five I celebrated my first Christmas with Ronald. He gave me enviable presents—a diamond ring and a real mink coat. Any *other* girl would have been happy to receive such tokens of generosity, but then again, I wasn't just any other girl—and my happiness was contingent upon things being right in the home. And they just weren't right. In fact, they were wrong, wrong and wrong. Ronald was selling and using drugs. A poster child for all that I called "unholy" I never got used to him.

Not entirely.

Ronald came from a large family and he specialized in "running game." His father owned a famous club in Cleveland called The Canteen and Ronald had a number of Speakeasies around town. I liked Ronald's family about as much as I liked him, which was not at all!

I wanted out.

Ronald was always doing grown folks stuff, and that created an environment that was no place for a kid to grow up.

Ronald drove a wedge between me and my mother. While she was busy catering to him and his needs, my brother and I were on standby waiting on our turn—which never seemed to come. Soon, I began to resent her. Our relationship began to come undone—which was easy because it wasn't wrapped too tight to begin with.

"You need to find us a father," I said in no uncertain terms to my mother one day. Imagine the look on her face when I said that.

"You have a father," she insisted.

"He ain't here!"

"Well now you got another one."

"I don't like him!" I retorted.

"Too bad," she said with her arms folded over her breasts. *Too bad* were the magical words that big people used when they didn't know what else to say. "Get over it" was also a reliable standby.

Too bad.

Too sad.

Get over it.

I wish I could . . . even still to this day.

There wasn't much relief or redemption in the choices my mother made—so I had to "get over it" enough to move on with my life.

In 1963 Martin Luther King was dreaming out loud that America would live up to its true creed that "all men are created equal." In many ways, I was also dreaming the same dream. I did what I was told and conformed to what was expected of me. I did it in joy expecting the payback of a "good life." One filled with promise and high achievement.

On November 22, 1963 John F. Kennedy was assassinated in Dallas Texas. Somebody literally blew the back of his head off as he drove through Dealey Plaza with his wife by his side.

I was at my grandmother's house when the breaking news appeared on the television set. I can remember everybody in the house crying. I was just a little girl at the time, but found myself overwhelmed by the sadness in the house. *Should I be crying too?* I wondered. Everybody seemed so sad, but they were sad beyond reasons that were greater than my understanding at the time.

America's 35th president, John F. Kennedy was dead at 46.

Kiss Civil Rights goodbye.

We are screwed again.

That's why they were really crying.

Tears, tears and more tears.

In 1964, we moved to an uppity neighborhood in Shaker Heights. The area was scarce on Black folks. There weren't a lot of us around, and with the exception of football player Jim Brown, who lived around the corner from us—it would have been difficult to find people of color in this upscale part of town. Moving here was another major adjustment in my young life but thank God for kindergarten! Once I started school, I found it to be a welcomed relief. The classroom pulled me out of the grown-up world of Ronald and Mary and reconnected me with the heart of a child. I loved the atmosphere of learning and soaked it up like a fresh sponge—bone dry and hungry for knowledge. "Welcome to kindergarten, Cheryl," said my first teacher, "you're going to like it here very much."

I was a dedicated student and a rather quick study. My intellect came in handy in the classroom and it was an invaluable asset in the *real* world. And just what did a six year old girl know about the real world? I knew enough to tag along on the heels of my uncle Willie as he worked the rigors of his own electrical business.

"I wanna learn . . ." I insisted as I eye-balled his every move. He was meticulous and he was intelligent. I fell in love with both qualities and wanted to emulate them—*for not a single day of my life had I ever been an ordinary chick.*

"What do you wanna learn, Cheryl?"

"I wanna learn about electricity," I said with my hands on my hips.

"Well," he said standing up straight, "don't stick your fingers in no light sockets!"

"I already know that!" I said running behind him.

"Then you're smarter than your little brother!" he bellowed with a laugh.

"Teach me what you do . . ." I insisted, following him from room to room. "I wanna be just like you."

"You can't be like me," he said with a sharp tongue, "this ain't no job for a girl! This is a man's job!"

"You make your own money," I said. "Why can't I be like that?"

"You can . . ." he said stumbling around his words trying to find the *right* thing to say.

Uncle Willie was nobody's fool and the wrong thing said at the "right" time could land him in a whole heap of trouble and he knew it. I was impressionable, strong-willed and stubborn. All things combined made for an explosive combination. Uncle Willie knew that I took things deep into my soul and he didn't want to impress my heart with anything that might come back and bite him in the backside.

"Don't you want me to make my own money?" I further questioned him.

"Sure I do."

"Then why won't you teach me?" I pressed.

"Okay," he said giving in after a few beats of gigantic hesitation. "I'll show you how to run wire . . . you wanna learn how to run wire?"

"Yeah!" I said, answering quicker than he could ask the question.

And that's how it all started.

At the age of six I was running wire. Little did I know in that moment that I was being prepared for a life I had no idea that I would one day live. Sometimes we have to remain in the "dark" on our future—lest we run straight out of the pages of our own drama screaming that all hell has just broken loose and poured itself out at our feet in the form of our "life story." Interestingly, around the time I was learning how to run my first wire the Civil Rights Act of 1964 was passed which prohibited discrimination on the basis of race, color or national origin in programs that were federally funded. This included "women" as minorities; therefore, women were getting liberated too.

A "man's world" was about to change.

Here I come!!

The fighters of the era were emerging and I was shocked to discover that I was one of those fighters. It was during this time that I found a personal hero in Mohammad Ali. As a young girl, I never missed any of his fights and would claw my way to the front of the television set, amidst boys and men alike, to claim my rightful seat. Enchanted by his unyielding confidence and the way he moved—Ali seemed to walk just "above" water as he hypnotized both the crowd and opponent alike.

"He's a fighter and a winner!" I thought to myself every time I saw his reflection. Through the inner conviction of my spirit, I, too, had always felt like a fighter, but alone that wasn't enough. I also wanted to be a winner.

"I want to be a winner like Ali," I boasted aloud.

"Yeah baby . . . sure you do!" the men would say to appease me. But I meant it. By first grade, I was sold on the hype of winning.

Money.

Power.

Respect.

And shiny things that glowed in both the dark and the light.

Yes, winning brought *good things to life.* And I was bound and determined on having those good things. It was the good things that came along in the form of my dreams. I dreamt of the good life. I dreamt of King's speeches and ideologies and I contrasted it against another movement on the rise—the Nation of Islam who had a much more aggressive answer to Dr. King's non-violent approach. Both philosophies were mixing inside of me—juggling me around for sport. I felt like I was in the ring myself—but unlike Ali, I was unclear as to who the opponent was and what sort of fighting style I should be using.

Just fight, Cheryl, I told myself. *You'll figure it out someday.*

While Ali was fighting abroad, Mary and Ronald were fighting in Cleveland. The only difference was there was no substantial "purse" for participants and at the end of the day everybody lost.

I lost respect for my mother.

She lost respect for herself.

And I don't think Ronald ever had a damn bit of respect to begin with. Docile and unnervingly accommodating, I couldn't help but feel sorry for my mother . . . and from that deep well of sorrow grew compassion.

She couldn't help herself at all, I reasoned.

"I'm good and sick of you, Mary Waiters!" shouted Ronald.

"Why are you being so difficult?" she would ask.

But there were no answers—only more ass whoopings. On one particularly merciless day, I tried to jump in between the two of them and save my mother by attacking Ronald with my baby doll. Well, needless to say my efforts were unsuccessful. She got her butt beat anyway.

I tried Mama . . . really I tried.

During this time, my father had gone MIA and there was nobody to cry to. I couldn't cry to Mary because she couldn't hear me. She was too busy drowning in her own tears over Ronald—who from where I stood was as sorry as the day was long. And though he had money, he didn't hold a candle to being the measure of a man I had hoped to find in a father. In fact, meeting Ronald made me want to be a better *man* than he ever thought about being; and ironically, I would spend most of my life doing just that.

Chapter 5

"All changes, even the most longed for, have their melancholy; for what we leave behind us is a part of ourselves; we must die to one life before we can enter another."

Anatole France

Life in Shaker Heights was pretty on the outside, but undeniably raw on the inside. And just like a newborn babe, I breathed in the air of an upscale world—all the while trying to make sense out of it. The Heights were a drastic departure from our previous neighborhood in Cleveland. There were white people everywhere, and I was in awe of the new color scheme.

Had I ever seen such a high concentration of white folks in such a small and intimate space? The length of a city block or two, maybe even three.

Welcome to my new world.

Again.

It was a sea of floating salt and I was the pepper sprinkle in the middle of it—going against the grain of acceptable race relations. But what did I know? My eyes were stuck on "wide open."

Taking it all in.

Soaking it all up.

Maybe it was less the presence of white people and more the absence of black people that I found so captivating. So deeply mesmerizing. Pulling me to the edge of my comfort zone and pressing me away from all that I had known prior to my arrival at this exquisitely crafted and beautiful home.

The Waiters had just been upgraded and within no time, I was back to the business of being a kid again. I sought out new playmates in Shaker Heights and I found them.

Boys.

I couldn't seem to get away from a male dominated world.

Okay.

Boys it is.

They were the sons of NFL football player, Jim Brown.

They were nice enough, I guess. They were boys so I didn't expect much by way of courteous behavior. I used to play with them religiously, not because I was so crazy about them. It was just a good way to pass the time. I didn't mind playing with them but I did feel a bit of angst when it came to their daddy, Mr. Jim Brown. He was a colossal-sized man with a strong voice and a daunting presence. I never did like the way he treated his wife. He abused her. He had an internal aggression about him that I never understood. Mr. Brown's behavior reminded me of Ronald and I didn't like it one little bit.

Abuse terrified me.

I was never abused directly, however, observing Ronald's temper and frequent outbursts of anger set the stage for the bitter taste that violence always left. It was an unwelcomed expression in my world and I wanted nothing to do with any of its associates. And as far as I was concerned, Jim Brown was a close associate. Therefore, I found myself ducking and dodging Mr. Brown. I could also feel myself judging him for being a bully.

Mean Bastard, I wanted to say.

But being in the NFL seemed to provide him with a prestige that seemed to skate just a hair above the law—at least in the way he treated women, mainly his wife. People looked up to Jim Brown.

He was somebody.

He had a reputation.

He had things.

He had money.

He had everything or nothing at all.

He could have been one of Ronald's idols.

Who's to say?

It was here in Shaker Heights that our family started mixing it up. Ronald and Mary started hobnobbing with the stars. Well, maybe they weren't really stars, but they were as close to celebrity as the city of Cleveland had to offer. Anybody who was "somebody" came to party at our house. And that experience was always more of the same—drinking, gambling and swearing. I used to hide behind the couch and watch in awe at all of the shaking booties and gyrating hips.

The cursing.

Booze.

Smoking.

And gambling.

It was all grown folks stuff.

Grown folks stuff is what eventually got Ronald into trouble . . . and I mean *big* trouble. He was dealing with the Devil and the demons that he stirred up came back to haunt him. Ronald had double crossed the Mafia and it was a "pay day" in hell for sure.

Time to get a move on . . .

Wait a minute . . .

Hold up . . .

Just when I was starting to like Shaker Heights—the neighborhood.

The school.

My friends.

The sounds of Motown were everywhere.

Black folks seemed to be getting a "leg up" maybe even two.

Life was turning around.

We began to believe again in the promise of freedom.

Hope was on the rise.

Life was on its way to the "good" side of things—but before I could even ask, "What's going on?" Mary was throwing clothes in a suitcase and Ronald was running the bags out to the car so quick that it made my head spin.

"Come on! Come on! Get a move on!" shouted Ronald.

"Come on kids! We gotta go!" said Mary.

"Where we going?" I asked.

"Don't you worry 'bout that . . . just get in the car!"

"But I wanna know . . ."

"You don't need to know everything," scolded Mary.

"Not everything," I said, "Just *one* thing . . . where are we going?"

"You heard your mother!" snapped Ronald. "Get in the car . . . stop being nosey. You'll see where we going when we get there."

"I'd like to know *before* we get there."

"You so stupid!" said my brother, Chuckie, taking a swing at the back of my head. "Just get in the car."

"I don't like this one little bit . . ." I said with my hands on my hips as I walked just as slow as I possibly could to the car.

"You ain't got to like it . . ." said my mother.

Chuckie laughed and pointed . . . made some funny faces and even stuck out his tongue. I didn't pay him any attention at all—just got in the car and surrendered to the interruption of *everything*—yet again in my life.

The sign said welcome to Springfield, Ohio. It may as well have said, "Welcome to Whiteville USA." I think Springfield may have had two more white people than Shaker Heights did. Situated on the Mad River, Springfield was a little town approximately 45 miles west of Columbus and 25 miles northeast of Dayton.

Our family took up what I referred to as "temporary" boarding with some of Ronald's extended family. We had to live "low key" because Ronald was on the run from the mob—though none of us knew that at the time. Honestly, how do you explain that in casual conversation at the dinner table? "Yeah . . . Mary . . . pass me the potatoes . . . by the way, we had to get the hell out of Dodge because I stole some money from the Mafia and they're looking to cut my throat. Got any dessert?"

No, that would have been breathtaking, heart-stopping conversation. Instead he opted for a less threatening conversation, "Ya'll . . . meet my Auntie . . . this is your new home. Hope you like it."

Springfield was a simple place and this was a simple town. The residents led simple and uncomplicated lives. This made me wonder if we ever fit in. We didn't feel so simple to me. Ronald felt dark and complex and Mary, neutral and undecided.

I never understood it.

Mary was devoted to the Bible and for whatever the reason, she thought that being a good Christian was synonymous with living in poverty. I, however, saw things from an entirely different perspective. It seemed to me that everybody God dealt with became rich. Biblical heroes David and Solomon were rich, as well as many other people from the Bible—so I naturally assumed that God wanted me to be rich too. I mean . . . honestly . . . if we're going to make an error in interpretation—can't we error on the side of us having deep pockets?

In addition to a poverty mindset, Mary also had some kind of bizarre fixation with "Ronald the hustler" and in the moment, I imagined that she would follow him to the ends of the great Earth—or worse yet . . . even further. I never understood my mother's obsession with Ronald. By my account, he was a "no good for nothing . . ." But let her tell it—he was "to die for." She seemed far less interested in bettering herself by way of education and more concerned about "snagging" a hustler to help take care of her and her children.

Mama, can we look into getting you a trade?

She didn't belong with Ronald any more than we belonged in Springfield.

Everything in Springfield was borderline generic, at least in my mind. None of it felt permanent and it never felt like home. It just felt like some kind of place that I was staying in the meantime—in between time—waiting on the rest of my life to arrive. But that was tricky because it never seemed to get "here or there" quick enough. And while I waited on "change" that never seemed to come, I would ease my suffering through books.

"I like to read," I told my kindergarten teacher.

"Reading opens your mind, Cheryl," she said to me.

"I want to have an open mind," I would respond with a big smile.

"You're going to go far in life . . ." she promised.

"That sounds real good to me."

"Always do your studies and do what you're told," she asserted.

"Yes Ma'am . . ." I replied without question.

"You're a good citizen, Cheryl," commented one of my teachers. I never forgot those words.

A good citizen?

Sounded like a lot of responsibility to me.

Going far sounded like a viable plan, so I was eager to do whatever it took to get to wherever I needed to be. I wanted to go much farther than my mother had gone. I'd lose myself in books—reading, writing and critical thinking. As a child, I had an affinity for abstract concepts and mathematics. I wasn't like a regular kid. I was more like an old woman in a little girl's body. Not only did I have questions but I also had the answers. It made for an interesting paradox of a young black girl growing up in the 60's in America when all hell was breaking loose—and we were one step from freedom and another step from a cage.

But what did I know between kindergarten and first grade?

Nothing *and* everything.

I was living the American Dream, or so I thought. Everywhere I went I was surrounded by white buildings, white fences, white snow in the wintertime, and Snow White in the summertime; and at the end of the rainbow I wanted to be Cinderella.

In Springfield I was introduced to the American way of life. I was Pledging Allegiance to the flag of the United States of America and singing "God Bless America" every chance I got. And in between beats of that harmony—I didn't mind boasting:

> *This Land is Your Land*
> *This Land is My Land*
> *From California, to the New York Island*
> *From the redwood forest, to the Gulf stream waters*
> *This land was made for you and me.*

I was being groomed to become a proud American citizen.

Land of the free.

And home of the brave.

I lapped it up and didn't leave a crumb on the table. Whatever I was told, I believed without question. I followed the rules—every last one of them. An avid student, I was a zealous participant in every forum of learning. I was patriotic all the way down to the red of my blood and the white of my bones. And by now, I was loving white people so much that I really wanted to be one.

In short, I was a sucker.

The energy of the era was pulsating through my veins and charting a course in my own personal destiny. It was forming an outline of the world yet to come—I just couldn't see it yet.

My destiny.

The Women's Movement of the 60's was gaining momentum and riding on the tail winds of the Civil Rights Movement. The spirit of rebellion was resting at everybody's feet. No wonder we couldn't sit still. The Sexual Revolution of the 60's—which was spearheaded by the creation of the Birth Control pill—was dancing all around me. It was changing everything—right down to my own heartbeat. *Hold on to your "hip huggers" and "Peace signs" . . . things are about to get a little crazy.*

Chapter 6

"She was a girl who knew how to be happy even when she was sad. And that's important . . . you know."

Marilyn Monroe

On May 18, 1964 Mary Waiters went and did it again—but this time her partner-in-crime was Ronald. Initially, I wrote this line in "jest" but in reflection on Ronald's character—this may be less of a pun that I had intended. After all, he was a real life gangster. However, I wasn't referring to his criminal activities this go around. I was introducing him as a "partner in parenting" as Ronald and Mary brought a new addition to the family—a baby boy who would go by the first name of Michael and the last name of Gilmore, which was Ronald's last name.

The birth of my baby brother was not a widely celebrated affair. My father, Charles Waiters, probably didn't take too kindly to my mother reproducing with another man. But then again, I don't know when he would have had time to notice that she had gone on with her life because he had long since moved on with his own. Nevertheless, shortly after Michael's birth, Charles made it official by filing for divorce on June 5, 1964.

The thrill was gone.

This was the birth of Mary's third child and her make-shift, built-upon-sand and barely holding-itself-together relationship with Ronald was a little shaky. Or maybe it was a lot shaky as my mother struggled to keep the dysfunction of the relationship hidden from plain sight. But secrets only lay dormant for so long before erupting because everybody knows that every dark night eventually crawls its way into the light of day.

Michael's birth was telling—in more ways than one could possibly imagine. He was born a *blue baby.* The real life infant photo of a blue baby is nothing nice. Difficult to care for—he was born with complications from his father's drug addiction. He had seizures, a brain wave malfunction and heart failure. And ultimately, when nobody's touch could bring this little baby comfort, everybody turned to the mother and asked, "What the hell?"

"I never did any drugs . . ." swore Mary.

Then what's wrong with this baby?

"I said I ain't never did no drugs!"

Then why won't this baby stop shaking?

"Ronald," was all she could whisper with her head hung low. "It was Ronald."

Can we leave him now? I felt like asking but it was no place for a child to question an adult with real life grown-up questions that little kids didn't have permission to be asking much less know the answers to. Needless to say, the bare facts of the situation and my ability to address the drama as a "youngster" left me stuck between a rock and a hard place—and somewhere else between a lament and a lullaby.

So where do we go from here?

Eventually baby Michael got over the withdrawals and life went back to normal again. But I always add my disclaimer here: *whatever normal was*—because when the dust finally settled I was out numbered two to one.

Two boys.

One girl.

The birth of my youngest brother marked the beginning of the rocky road down the path of "gender differences." At family gatherings and just about everywhere we went there was a notable distinction between the "sons" and the "daughter."

Chuckie.

Michael.

Ooooh Laaa Laaa.

The boys got all the gooey stuff: the slippery affections of the aunties, grandmothers, cousins and even our very own mother.

Chuckie.

Michael.

Ooooh Laaa Laaa.

I got the leftovers.

Cher and the boys.

It became a standardized greeting in our household. It was used by everyone. I was distinctly separate. I lived in my own zip code within our home. I had a special permit to reside in what I refer to as a "specialized" environment for girls only in an all boy world. An outcast—sent to live out the remainder of my days on my own Continent by virtue of the fact that I bore a vagina. Mother Nature—why would you play such a cruel trick on a girl and carve her into the female gender when she wasn't looking, only to place her in a hostile environment filled with ignorant males? The adult men around me, particularly the friends of my father, categorically referred to me as "the girl child."

Hey girl.

Little girl.

Big girl.

Eventually I bled all the way out until I became nameless.

Just girl.

Girl.

"I need you to watch after your brothers," my mother told me when she went back to work after Michael was born.

"I need you to be responsible," she added.

"Okay," I said.

"Make sure they eat and that Michael's diapers are changed."

"Okay," I agreed again.

"I really need your help, Cheryl . . ." she insisted. "I got to go back to work and we gotta make this situation work out all right."

Lost for words, now I offered only nods.

"You listening to me, girl? I need help . . ."

Girl?

Girl?

"Yes ma'am."

"I'm running a tight ship here and there ain't no room for loafing."

"No loafing here, ma'am."

My mother was a perfectionist and that was one of her highly disturbing qualities, I might add.

"These boys need tending to . . ." she reminded me constantly. "*Girl* . . . you hear me?"

I wasn't a grown up yet, but I sure did feel like one. Not sure how I got suckered into it all—but I started feeling like the mother and Mary started feeling like the child. Once she saw that I was responsible then the rest came easy.

Just let Cheryl be responsible.

Mary took up the business of working and having a life again.

"Uhhh . . . excuse me . . . but ain't these *your* kids?"

Well . . .

Obviously that comment wouldn't have gone over well so I left it unsaid. The words dropped straight from my head to my heart and there I let them rest.

Perhaps.

By this time there was an unwritten rule in our home. It went a little something like this: You will not like your mother and rarely (if ever) will you get along with her.

Ever again.

Never again.

It was almost like a curse.

A mother/daughter prophesy of doom.

"Uhhh . . . excuse me again . . . but ain't these *your* kids?"

These "kids" and me were on our way to real honest-to-God sibling rivalry. And yet still . . . everybody was loving on the boys. However, that's not to say that some of the male relatives showed me favor but the bulk of all the "good stuff" was dished out to Chuckie and Michael.

Dang.

I didn't want to be a boy but I damn sure wasn't a big fan of staying a girl.

Little girl.

Big girl.

Smart girl.

Pretty girl.

Whatever kind of girl.

It was all overrated.

Now being a "boy . . ." that's where the benefits came from.

Eventually, I gave up on relationships with "real" people and adopted a make-believe friend. Well, he was sort of make-believe in that he *was* real and he wasn't real all at the same time. His name was the Great Gazoo, and I met him while watching an episode of the Fred Flinstone show. Gazoo was a green, futuristic alien who was banned to Earth in the middle of the Stone Ages after he was kicked off his planet for creating a doomsday machine that could potentially blow up the whole Universe.

My kind of guy.

I was always a sucker for a boy with half a brain.

On the television show, only Fred, Barney and the kids could see Gazoo. Wilma and Betty couldn't see him because they didn't believe in him. I didn't give two shakes what Wilma and Betty had going on cause I believed in Gazoo with all of my heart, and more importantly, he believed in me. He hung out with me and made for good company—especially when I could have easily surrendered my attention to boredom.

"You think I'm pretty, Gazoo?" I asked him one day while in the middle of scrubbing the tub.

Yep, he answered without hesitation.

"I do too . . ." I said with a nod. "And what about smart?"

The smartest.

"Me and you think alike . . . I like that Gazoo . . ."

Me too.

"So why can't these knuckleheads around this house see how special I am?"

Dum Dums.

And I'd laugh and laugh.

"They are dum dums aren't they?"

Sure are.

"When I grow up I'm going to be somebody real important?"

You're already important.

"But more important than I am now. You believe in me, right Gazoo?"

As much as you believe in me.

"I believe in you a lot . . . and just cause can't nobody else see you . . . it don't mean a thing to me."

And Gazoo would laugh out loud.

"Sssshhh . . ." I'd always have to remind him, "just cause they can't see you don't mean they can't hear you."

Right . . . I forget sometimes.

"Gazoo . . . you know what I was wondering . . . why are you so green???"

The same reason you're brown, I guess. It's the way God made me.

"You ever feel left out, Gazoo?"

Sometimes.

"Me too . . . that's why I'm glad we got each other."

"Who are you talking to you weirdo???" Chuckie would shout, rudely interrupting my conversation with Gazoo to accuse me of unpleasant things . . . like insanity.

"Mama . . . Cheryl's talking to herself!"

"Get outta here and leave me alone, Chuckie!"

"Mama . . . Cheryl's a weirdo!"

Weirdo?

Not at all.

Gazoo had other friends besides me . . . he had Fred, Barney and their kids as confidants. And on top of all of it . . . he had his own television show which is more than any of those little losers around me had. So, I did not have a problem at all with Gazoo being my best friend. In fact, I would have been *crazy* not to be his friend. But not everybody saw it that way. Gazoo and I hadn't even been friends that long before my mother suggested charm school.

"Charm school?" I said with my face twisted into a permanent frown.

"Charm school," she confirmed.

"I'm already charming," I protested all the way there.

"Ha! Ha! Ha! Cheryl's going to charm school!" Chuckie teased and teased.

Damn.

Damn.

I think that was one of the first rules of etiquette—ladies don't swear, but neither did I . . . yet anyways. I was only a little girl.

Charm school, like every other course of study that I was introduced to, was a breeze for me. And like everything else that I was taught while growing up—I took it far, deep and wide into my soul. In learning the rigors of poise, confidence, etiquette, manners, grooming and the elements of fantastic conversation—I emerged from charm school thinking I was a mother loving movie star.

"Oh brother . . ." said Chuckie. "Mama . . . Cheryl thinks she's a beauty queen now!"

"You're just jealous cause you ain't as pretty as me . . ." I said to him, "but don't worry . . . someday a beautiful princess will come along and kiss you . . . and change you from an ugly toad into a respectable *good citizen.*"

"Mama . . . Cheryl's still a weirdo!" he ran through the hallways shouting.

What a strange little boy he was.

Charm school may have been charming, and I may have learned correct posture—but that sure didn't ease the burden of being the only girl in an all-boy household. At every turn of the twisted road there were painful reminders that I was indeed the strange one—unique unto herself—a hybrid being born without a penis.

Why God?

Why?

Feminine creatures were stomped upon in my household. Cruelty to the female species was condoned in every part of life and perpetrated through the manipulation of "innocent" television shows that doubled as a two edge sword designed to wear us down from the inside out.

The Little Rascals, an acclaimed 1950's show about a group of poor kids from the neighborhood and their wild adventures, even got in on the action when they released an episode where Alfalfa and the Gang created a club called the "Women's Hater Club."

I vividly recall this episode where they put all of the girls into the "Women's Hater's Club." Now, the problem with this was my knucklehead brothers were watching this episode and thought that this sounded like a "real good idea" and guess who was the first girl they introduced to their new, pitiful club?

Me.

And with a "hard slap in the back of the head" from Chuckie, I was inducted into their hall of fame. He sealed the deal by yanking my ponytail as hard as he possibly could.

Yikes!

"I'm going to tell Mama!" I screamed as I ran out of the room. But telling never helped and with each passing day my brothers aggravated me more and more—with constant "love taps" in the form of aggressive aggravation.

"Can you please give them back to God?" I asked my mother.

No can do.

God gave them to you.

They're your bothers.

In all fairness, my mother didn't have time to keep tabs on my wars with Chuckie, because she was raging plenty of her own with Ronald. There was a violent blow up between the two of them when I was seven years old. It was the worst ever and I thought for sure this was it—but damned if it was. They still stayed together. They stayed till Ronald eventually came to his senses and decided it was time to go back home.

"I can't keep running," he said. "I'm going to turn myself in."

"What?" Mary asked devastated.

"This ain't no kind of life . . . hiding out, ducking and dodging . . . from everything and everyone."

"So what's going to happen to us?" Mary wanted to know.

"We're going home," he confirmed.

And just like that we were en route to Cleveland and back with my mother's family.

It was like we never left, so in many ways it felt like we should be leaving and going someplace else. Some place that we could call our own, a real home. I was tired of living like a village Nomad—always occupying somebody *else's* space.

Anger was breeding in the air and reproducing in like form. In the summer of 1966 race riots broke out in Cleveland and lasted for six nights straight. They were an off-shoot of the Watts riots in Los Angeles. Four people were killed and more than 30 were critically injured. There were some 240 separate fires set throughout the city.

And why?

Damn fools is what they were. People burned up the very streets that they themselves walked down. So, where do we stand now that we've run out of room?

In my grandmother's house the air was not much different than "riot" air as my mother served up the "last supper" for Ronald. I showed up at the breakfast table mad. By eight years of age, I had the "anger act" down cold as I sat at the table that morning with body language which suggested that everybody at the table could kiss my little yellow behind.

In truth, I was mad about everything.

Mad about breakfast.

Mad about moving out of Cleveland and mad about moving back.

Mad that if you rubbed two pennies together, my own father still wouldn't have been worth two cents.

Mad about Ronald.

Mad about Mary.

And mad about life.

Well, that pretty much covers it all. Ronald was about to ship off for an eight year term in the "Big House," and I can honestly say that I wasn't the least bit mad about that.

Farewell.

And Goodbye.

On the other hand, Mary was sad about it—desperately sad—and it seemed as though her tears bled all over the hot butter biscuits and pancakes. She prepared a feast fit for a king and served it to a loser while they made small talk.

"Did you sleep well?" she asked him.

"Slept fine."

"You want some more pancakes?"

"I'm good," he grunted.

"Grits?"

"Pass 'em down," he said.

While they were passing grits and hotcakes around, smoke was fuming from the inside of my nose.

Hey! I wanted to shout.

I am right in the middle of my life and it ain't going so hot!

Mama, can I get your undivided attention right about now?

"Cheryl . . . did you say something?" my mother asked, looking up with a surprised look on her face.

"Huh?" I gulped. I hadn't said a word but maybe she could read my mind.

After breakfast the rest came easy. Ronald turned himself in to the authorities. He was wanted both by the mob and the police, so he figured jail was an easier sentence than a mob payback. He packed eight years worth of clothes that morning, and then, he up and disappeared right out of our lives leaving my mother with a two year old son and a whole lot of bad memories.

It was time to start over again. I just didn't realize that we would be starting over so early in the morning. Not long after Ronald shipped out, my mother who had taken on a part-time gig as a "bar maid" in the evenings, met a man who would change her life forever.

And ours.

Again.

His name was Thomas Strickland.

Mary took his breath away upon first sight and he made her an offer she couldn't refuse.

"Come and live with me in Buffalo, New York," he offered.

Sure.

Why not?

Ain't got nothing else better to do tonight.

It sounds like a joke, right?

It wasn't.

A few hours later, my mother picked me and my two brothers up from the babysitter and we boarded a bus leaving Cleveland. It was 2 o'clock in the morning and we were on the move *again.*

Chapter 7

"I love them and they answer me with pain and torment."

King Arthur in Camelot

At 2 a.m. you don't ask a lot of questions even though you *have* a lot of questions.

"Where are we going?"

"Who are we going to live with?"

"And what are we going to do when we get there?"

"Can Gazoo come too or do I have to sit this one out alone?"

There were no answers—only the ricochet effect of hollow thoughts bouncing around in my brain searching for a place to plug into the real world. Again, there was no dialogue of explanation from my mother and no firm conclusions based upon the facts. And just what were the facts?

Fact #1: We had moved before.

Fact #2: We were moving *again.*

Those were pretty much the facts.

There was no "connecting the dots" of all the moving pieces.

No foundation.

No sketch pad with a thoughtful drawing already in progress. Everything was done freehand—which meant it was scattered and disconnected. This was the epitome of what you called an adlib. We were making up this whole damn thing known as the "story of our lives" right in the middle of it—at least that's what it looked like Mary was doing. All we had was a couple of suitcases and our whole life savings—which could fit into the palm of *one* of my mother's petite hands.

Now that's what you call improvising . . . for real.

There was just the big black sky hovering overhead with no promise of dawn to come, and the steady crawl of the Greyhound bus as it rolled down the highway whereupon Cleveland faded out and Buffalo faded in.

Good morning America.

Welcome to my world again.

Where the heck do we go from here?

Thomas Strickland was a forty-year-old steel mill worker who from the moment he laid eyes on Mary Waiters had an undying affection for her. She took his breath, his heart and claimed a small piece of his one bedroom cottage for us to lay our head down at night.

I was impressed with Mr. Strickland. A self-educated man—he resembled the actor Ossie Davis and boasted sparkling eyes overflowing with laughter. His home was filled with books and adorned with all things beautiful. I felt like asking, "Mama . . . did we just move to Camelot?"

Thomas had a sharp mind and an affinity for learning. He wasn't allergic to hard work and earned his wages on the "right" side of the law. This alone brought him into my good graces fairly early on. I had no quarrel with him. At first glance, he seemed to be a respectable man with honorable intentions. His essence sparkled with the untold tales of a gentle man and after dealing with Ronald and all of his drama for so long—I was in a ripe mood for *kindness.*

"I like him," I said to my mother shortly after our arrival.

"He's a'ight . . ." said Chuckie . . . not really caring too much either way.

"We gonna stay here?" I asked hoping that she would say yes.

"This is our new home," said my mother.

I raised my brow.

Wow, I thought to myself. *Mama ain't no fool.*

"I want to talk to you kids," Thomas said, interrupting my mind chatter.

Uh oh.

I hope Chuckie ain't broke nothing in this man's house. We've haven't been here ten good minutes yet.

He seated us on his sofa and sat directly across from us. His forehead seemed to crinkle under the weight of something serious. Would his words puncture Chuckie and me, leaving us for dead as we sat upon his couch? I swallowed hard at least two times because this meeting of the minds was looking more and more official by the minute.

"Yes Mr. Strickland?" I inquired humbly.

"Call me Thomas," he said with a wide and friendly smile.

His easy demeanor was non-threatening so I relaxed and leaned back. Chuckie dropped his shoulders and eased back a bit too. Michael was only two years old and he didn't care one way or the other, as long as he got fed on time.

Yeah, I liked this Thomas Strickland guy.

"I know you kids have been through a lot . . ." he said with a calm voice and steady eyes. "Your mama's told me some things . . . but it's all gonna be different now."

Silence.

I didn't respond right away and neither did Chuckie. Frozen in time or perhaps just outside of it—we only made brief eye contact and then quickly returned our gaze to Mr. Strickland . . . I mean, Thomas, lest he think that we were being rude by not accepting his hospitality.

"You hear what I'm saying?" he asked trying to get a feel on our emotions.

Chuckie's eyes widened and he nodded out loud.

I followed suit.

"I will never hit your mother . . ." he promised, "and I want to adopt you kids."

"But . . ." Chuckie started to interrupt until I poked him in the side before he could finish and gave him *that* look. That look suggested to him that he should "be polite" and not get us thrown out of this man's house and into the freezing cold night.

"Sssshhh . . . Thomas is speaking," I scolded him.

Thomas laughed. "It's okay . . . what did you want to say?"

"Nothing," he said shaking his head back and forth quickly.

"I'm going to adopt you kids," he promised. "We're gonna be a real family."

What a deal!

We accept!

We'll take the tall black man with the white picket fence surrounding his cottage home, a steady paycheck and the friendly face.

That sounds like a real good plan right about now.

Thomas had no children and we no longer had a father so it would seem to the "thinking man" that this was a match made in heaven. He immediately offered a peace offering in the form of a shoe box filled with fifty cent pieces.

Wow!

And he's rich too!

I was digging it—the whole situation from the scholarly-looking books on his shelf, to the beautiful sofa in his living room, and my shoebox full of shiny coins.

Lock stock and barrel.

I was in.

I'm going to have a real family again, I said to myself. *Maybe Gazoo won't need to come along after all.*

Chuckie and I took the sofa.

My mother, Thomas and Michael took the bedroom.

And this was the beginning of our storybook tale of Camelot.

We started going to church every Sunday and having lunch as a family after the service. *Things are really coming along,* I thought to myself. But before I could get one complete revolution around a single happy thought—it was interrupted by the sound of my mother's nagging voice.

"Cheryl . . . Cheryl . . . you listening to me?"

"Huh?"

"We need to talk . . ."

"We do?" I asked surprised.

"Listen . . . I'm the woman of the house," she said in that matronly tone.

What?

What?

I didn't know where this was coming from. I was blindsided by the "urgency" of Mother Mary's message. All of a sudden Mrs. Waiters was taking back the reigns of motherhood.

"I said I'm the mother of the house," she repeated, "not you!"

I was waiting for a drum roll or for the audience to stand up and cheer for the villain. This moment felt like a "bad bad ending" to a show that had offered promise from the onset.

What just happened here?

"I'm talking to you Cheryl . . ." she snapped.

The words stung as they rolled off her tongue and landed across my young, soft skin. They bit . . . and hard. I wanted to fling my head back as though I had whiplash because I certainly felt side-swiped by Mary's sudden and drastic change of heart. She never minded me being the mother in Cleveland? In fact, she insisted upon it even though I hadn't applied for the job and I never wanted it to begin with.

"It's all a show ain't it?" I asked her with attitude.

"What did you say to me?" she asked in a raised voice.

"Why do you always have to have a man???" I shot back with fire rolling off every word.

"Excuse me Cheryl . . ." said my mother with a stern voice and a defensive stand. "Would you like to continue . . . ???"

But I offered no reply because an answer would land me grounded and in a whole heap of new trouble. I already had enough issues with my mother, and though Mary Waiters may have duped Thomas Strickland, I surely knew better. He was a pleasant meal ticket but her restless and wild spirit would never take to the notion of settling down with a common man.

Why?

The answer was simple—she liked her men uncommon and dangerous. A blue collar wage earner would never be good enough for Mary Waiters, even if he could read and had a bit of culture and etiquette about himself. So she could drop the "Beaver Cleaver Stage Mother Act" at least with me because I knew the truth. But of course I played along—one . . . because I had too . . . and two . . . because I wanted the fantasy so bad that I just pretended it to be real. I was in the mood for a real home *and* a real family—even if the people posing in the production were as fake as a three dollar bill. I wanted to rest on ground that wouldn't come undone overnight if Mary changed her mind and took a fancy to moving on at daybreak.

It's not that I thought Thomas Strickland would leave us.

No.

I feared that Mary Waiters would never stay with him—for long anyways.

He wasn't her style and deep down inside I knew it. And I also knew that she wanted her style more than she wanted a family. She wanted what she wanted and it did not seem to matter the cost of her desires—and this was the basic conflict in my relationship with my mother.

Big sigh.

Or maybe it was just *another* to add to my growing list of unresolved complaints.

Well, nobody ever said Camelot was perfect did they?

I sucked it up and went on.

Mary took on the mother role and I took on other roles. I began to star in my own Off Broadway production called "puberty." One morning I woke up and I had lumps in my chest.

They frightened me.

What in the blankety blank is going on here . . . ????

I wanted to wake up Chuckie, who was still asleep and share this unusual discovery with him.

"What do you make of these?" I wanted to ask him, but then I reconsidered for two reasons:

1. He didn't have any "lumps" in his chest;
2. He was a certifiable knucklehead so I had to pass.

Mother Nature was calling on the inside and Mother Earth was calling on the outside. As a child I didn't know what to make of all that was happening around me, but I knew that something "magical" was going on. One day while I was out front playing in our yard, I happened upon a tree. It wasn't just any tree it was one that produced "magic" as far as I was concerned. My attention was immediately drawn to a single branch of the tree which stood out. It was a standout because the branch appeared to be made of cloth.

"That's interesting . . ." I said to myself aloud as I moved in closer.

Is this tree wearing clothes? I asked myself.

Out of the corner of my eye I saw Chuckie and Michael playing on the other side of the yard and I thought about calling Chuckie over to witness this discovery, but after careful consideration I opted out for one reason:

1) Chuckie was a certifiable knucklehead.

So I passed on sharing but not before spotting a unique looking cocoon growing on the branch. I was so drawn to it that I broke off that piece of the branch and took it in the house and put it in a Miracle Whip jar and hid it beneath the kitchen sink. I was fascinated by the discovery and monitored Mother Nature's process—both with my body and with this cocoon—till one day something amazing happened.

I checked under the sink and the cocoon had opened up and given birth to a giant moth butterfly!

"Oh my God!" I howled. "It's a miracle!"

It was the biggest butterfly I had ever seen in my life!!! The moment stood still and when it did—I took notes. *I'm just like that butterfly,* I said to myself.

"What are you talking about?" interrupted Chuckie, who was always barging in on my "ah ha" moments.

"Chuckie . . . look at this!" I said showing him the butterfly.

"Oh . . ." He said covering his mouth with his hand. "Mama's going to get you if she finds out you're keeping bugs under the sink!"

"Just be quiet and come over here and look at this . . ."

"I can see that big ole bug all the way from Cleveland . . . I don't need to move no closer."

"I'm this butterfly . . ." I insisted, staring in amazement at the jar.

"Mama . . . Cheryl thinks she's a bug now!" said Chuckie, running off and shouting down the halls.

He really was an odd child.

No matter what he or my mother would think about this moment . . . I knew that it was the beginning of something wonderful for me. I knew that this butterfly represented transformation in its purest form.

My transformation.

Change was surely coming.

It was quiet and yet it was so loud that the sound of the change was deafening.

It commanded my attention.

All of it.

No, it demanded it.

It was worthy of it and I was in awe.

Metamorphosis.

Defined as the "profound change in form from one stage to the next in the life history of an organism." It was then that I knew that my life would come to be exactly as I have read it described through this timeless quote by a source unknown: *Just when the caterpillar thought the world was over, it became a butterfly."*

And so did I.

Chapter 8

"May God put a spell on you . . . so you won't forget me."

Unknown

In the summer of 1967 Thomas Strickland made good on his promise of family. He married my mother in a simple ceremony at our home in Buffalo, New York. He also made good on the promise of adoption—beginning with my baby brother, Michael. But ultimately, Thomas would run into roadblocks in the attempt to adopt me and Chuckie. Charles Waiters was too proud to surrender his children legally—though he had already given us up physically, mentally and emotionally. Sometimes we hold on "in name only" and for all the wrong reasons:

We do it to save face.

We do it to nurture our stubborn pride.

Or perhaps Charles' unreasonable behavior was produced by the over-production of testosterone. And maybe at the end of the day, it was nobody's fault at all and we would have been the wiser to simply do as the song suggested: "Blame it on the rain."

After the wedding everything felt more official and I began to ease into the idea of security. I trusted family again. We had a real home and with each passing day I felt as though I was blending into something bigger than me—as the concept of family began to expand, breathe itself into life and become animated.

A mother.

A father.

Two brothers, albeit annoying.

Family.

Home life was good and getting better every day. Mary Waiters—who was now Mary Strickland—had up and done something good.

Something real good.

The props were phenomenal but the real show was just getting started. And in between the beats of life—there were always two things that I could count on: revolution *and* evolution. The only question remaining on the table would be, "Who was I destined to become when it was all said and done?" I always wanted to be on the receiving end of all good things, therefore, I did all that I could to draw inside the lines instead of outside of them. I thought there was real merit to being what is termed "a good girl." So, I tried to be it with all of my heart.

Good girl.

I went to church and looked for the Lord with an earnest heart. But all I seemed to find there was a lot of wild singing, shouting, and people giving themselves whiplash by falling out on the floor—all in the name of Jesus. And for a period of time, I went missing from the congregation.

"You're going to church today, missy . . ." my mother insisted.

"No I ain't . . ."

"Well you ain't staying home by yourself."

"I don't want to go," I pouted.

"Why?" inquired Mary.

"Because the people at church catch the Holy Ghost and I'm scared I might catch it too . . ."

"That's a good thing, Cheryl."

"No it ain't," I said shaking my head back and forth faster than Mary Strickland could talk me out of my bad attitude. I didn't want to be talked out of anything and back into church.

"Cheryl's scared of Jesus!" Chuckie went shouting through the hallways.

Did I ever mention that my brother Chuckie was really, really, really odd?

"None of the people at church even read the Bible . . ." I continued in my own defense.

"And?" Mary questioned rather loudly. In that moment I got the feeling that *she* was one of those people.

"So how do they know the Pastor's telling the truth?" I asked.

"Of course he's telling the truth . . . ministers don't lie."

Well, something told me right then and there that Mary Strickland was an unreliable source, because I had a feeling that ministers probably told more "half-truths" than regular folks, but I was just too young to prove it.

"I want to learn about the Bible," I insisted, "but I'm just not sure that church is the place to do it."

"That's blasphemous talk . . ." mother warned. "And you'll learn plenty if you just stop *thinking* so much."

The irony of that statement made me laugh, but ultimately, there were good and decent things that came out of our conversation. And before I could blow a single wish into the air—my theatrical drama debuted a new cast of characters who would eventually become my Godparents. Enter from stage left and into my life—Sam and Irene. They were a nice wholesome couple who lived upstairs from us.

Sam was a dark-skinned piano playing salesman and Irene was his high-yellow wife and the mother of his two children. They ran a very strict household and they did just about everything the exact opposite of our family. They were members of a faith known as Seventh Day Adventist. They observed the Sabbath on Saturday and ate special foods in observation of their Sabbath. Everything about them seemed right—or at least more right than what we had going on at the time. I was curious and wanted to learn more, and in due time I would satisfy my curiosity.

The Lord would see to that.

The Lord always does.

When we entered the Buffalo school system, my mother would have to fight to keep us current and in the same grade. For whatever the reason, the New York school system wanted to put us back a grade. This was "backsliding" at its finest, and they would have been successful, except my mother wasn't going for it . . . at all. In fact, she fought so hard to keep us current with our class, that I just knew she would be proud of my first report card which boasted of only the best—A's and B's.

"Look mama! Look! All A's and B's!"

"Yeah . . . yeah," she said putting me off, "you always get good grades."

In that moment my heart sunk.

Doesn't that deserve some praise or at least an acknowledgement? I wondered within myself.

She seemed complacent about the whole concept of grades until she read the horror of Chuckie's report card which boasted all D's and F's.

"Oh my God, son!" cried Mary, "Look at these awful grades! Terrible!"

Silence.

"With these kind of grades you're going to flunk right out of school!"

Chuckie shrunk in size at the outburst—then he began to cry.

"Come here son . . ." she said extending her hand to him in consolation, "I know you can do better than this."

She began to rub the top of his head and he was sucking it up like a newborn pup. I couldn't believe my eyes. I had watched Mary go from "furious to forgiveness" and from "mad to mothering" in seven seconds flat.

Are you kidding me?

Chuckie had done just awful and here he was getting all of the attention *and* all the love. Was he stealing my mother's love right out from under my nose? I couldn't help but resent him for it but no matter how much I resented him, I always resented her just a little bit *more*.

Didn't I matter at all?

Where was my hug?

Forget about the hug, how about a "handshake" for a job well done?

Guess not, I said to myself as I exited the stage for a costume change.

It was time to become somebody *better.*

Better would take some doing and it didn't come right away. In fact, for a period of time it would seem that things would get worse—both in my relationship with my mother and in the world-at-large. On April 4, 1968 on the balcony of the Lorraine Motel in Memphis, Tennessee—Martin Luther King was gunned down. In the gut wrenching moment of shock and disbelief, everyone was horrified that the Civil Rights Movement would be buried right alongside its leader. Non-violence and equality would have to find fresh soil to be replanted. In desperation and outrage people took to the streets and riots broke out all over the country. The dream seemed lost forever as I stood mesmerized in front of the television and taking in the news as best I could. I didn't know him personally but I felt the loss nonetheless. With Kennedy and King both gone—the world would be different now.

Strangely different.

But it would go on.

It always does.

And so would I.

In the summer of the same year, Chuckie and I returned to Cleveland for our first visit home since boarding the Greyhound bus at 2 a.m. in the morning. It was a summer of great change in my life and I was given glimpses of opportunity to make peace with the world-at-large and on a smaller scale—to make peace with my "tiny" world which seemed quite huge to me at the time.

I was fresh out of Buffalo with "attitude" towards my father—and reeling from the disappointment of who he had become or failed to be in my life. I openly poured my wounds on the table in a discussion with my Uncle Willie. As I sat in the kitchen and seething with resentment, I will never forget Uncle Willie's frozen face and penetrating eyes as he pierced with me with his words:

"Forget about your daddy, girl . . ."

"What?"

"You got a new daddy now. I hear he's a decent man so you just be a good girl and listen to him."

"Yeah but . . ."

"Ah . . . ah . . . ah . . . ain't no butt," he said stopping me short. "You listen to your new stepdaddy and do what them white people tell you do up there in Buffalo."

I stuttered.

Almost fell over—but I couldn't quite move my body enough to bend.

I was too numb.

And that was the end of the conversation—at least on that topic, so I had to find a way to "get over it" and "real quick."

Not.

I didn't take too kindly to Uncle Willie's advice, because honestly, it was riddled with error. But I did take to spending as much time as I could with my Uncle and learning the trade of electrician.

"Cheryl . . . Clean them steps off on the front porch. You hear me!" my grandmother insisted.

"Yes ma'am," I responded with one eye on those broken and busted up steps, and the other eye on my Uncle Willie as he packed up his bag and prepared for work.

"I want them steps shining . . ."

"Uh huh . . ." I offered my grandmother . . . knowing that was little more than appeasement because as soon as I got the chance—I was off those steps and jumping into Uncle Willie's truck to "stand in" as his assistant.

"What you doing, girl?"

"I'm going to be your helper today," I said smiling proudly. I was an outrageously confident ten year old.

"I don't need no helper . . ." he insisted.

"I work for cheap," I bargained. "I'll work for McDonald's."

"Cheap labor's good," he said with a nod, "You finish your chores?"

"Yes sir," I said, lying with confidence.

Uncle Willie smiled and we were on the way. I glanced back at those busted steps and knew my grandmother would figure out sooner or later that I had jumped ship on her assignment, and there might even be hell to pay for it—but it was worth the risk.

That day I spent with my Uncle Willie was nothing short of pure wonderment. In some ways, it changed my life forever . . . I just didn't know it at the time. Our first stop was a local store who had hired my Uncle to do some electrical work. I assisted Uncle Willie and followed his orders down to the last detail. When we left Uncle Willie was handed an envelope filled with money, bursting at the seams and overflowing. My eyes widened when I saw it.

Oh Laa Laa.

Cash is king.

I learned this at an early age. In fact, nothing said it quite like cash.

Our next stop was the private residence of a doctor. Again, I followed Willie's strict commands with precision that exceeded my ten years of age. Uncle Willie was impressed and I knew it—though he tried to be "professional" so that I wouldn't take any liberties as a "family member" and end up slacking on the job. And even though he was cool as a cucumber, he did glance my way several times—unable to contain his smile. And just for the heck of it—I smiled back.

Yep.

I was smart.

Real smart.

At the end of the doc's job, there was another exchange—which left Uncle Willie holding another envelope filled with money. My eyes widened and danced at the sight of the green dollar bills.

"I want a re-negotiation . . ." I said to Uncle Willie.

"Excuse me?" he asked stopping in his tracks.

"I don't want McDonald's . . ." I said boldly.

"No . . ."

"I want cash instead . . . I can buy my own lunch."

Uncle Willie laughed out loud.

"Well I'll be damned," he said. "You are your father's daughter after all."

Well, he was right about that. Getting paid was in my DNA.

Later that day, I met Jimmy. He was a shifty, on again/off again helper for Uncle Willie. He had some skills and knowledge, but he also had a pile of excuses as to why he was always late or a no show on Uncle Willie's jobs.

"Watch yourself," I said to him, "I just might take your job."

Jimmy thought it was funny and so did Uncle Willie.

"She's cute," said Jimmy with a smile—but I didn't smile because I wasn't kidding.

Later that day, Uncle Willie took me to McDonald's and with the fourty-dollars he paid me he paid me for my services, I was able to buy my own lunch. I sat amongst a table filled with electricians and ate my Happy Meal—while Uncle Willie's buddies talked 'girt and dirt.' It was shop-talk at its finest and I was loving it. Alas, I felt like I belonged to *something.*

I caught hell when we got home that evening because my grandmother's steps were still filthy dirty, but I was $40 dollars richer so I didn't mind one little bit.

Over the next week, I had to give up some "working days" with Willie in order to hang out with my father. When I saw Charles Waiters I was surprised that he even remembered my name. Maybe God had put a spell on him after all so that he couldn't forget me.

"Hey baby girl . . ." he said swooping me up in his strong manly arms.

Baby girl wasn't my name, but it would do for now.

"Hey my man . . ." he said showing Chuckie some love.

"Daddy!" said Chuckie with a smile. Chuckie always loved our father even if our daddy couldn't reciprocate the favor. As for me, I moved with caution in dealing with Charles, so I paced myself in walking to his shiny new car. It was a slow grind and not a hurried step. I had to move like that—because my heart was at stake. It had been stepped upon too many times by Charles Waiters without regard for the soft spots.

Charles had met and recently married a woman named Peggy. She was little more than a kind stranger. I had no connection or bond with her. She was a "stand in" model, almost like a Stepford Wife. She was just doing her job. Perhaps she was paid with a handsome salary to smile and be gracious with us. There could have even been a bonus package for not "beating the step kids." None of it was personal—and whether she liked or disliked us—nobody cared. She had accepted the part so now it was time to play the role. This was simply one slice of Peggy's "on the job" duties. It was though she had received detailed instructions from the Stepmother Manual. She had studied it well and I could find no fault with her performance.

Peggy and Charles lived in a nice home. They had beautiful clothes and ate good food. My father was still "all flash" and had plenty of cash from the local bar he owned *and* the games he was still running. Upon our reunion, I couldn't help but to compare him against the real men I had been spending all of my time with as of late—Tommy and Uncle Willie. And somehow, my father didn't measure up to the tall order of parent. It was written on his face and inscribed into his arrogant walk and flashy suits. He was charming and charismatic and had a pocket full of "bling" but something essential was missing from his heart.

And I knew it.

I could feel it.

He knew it too, and as a result, he made no apology for the knowing. And yet somehow . . . someway . . . this time . . . I wasn't as enamored with my father. But don't get it confused with my life story—you see . . . I hadn't gotten over him entirely . . . not by a long shot . . . I just loved him a *little* bit less than yesterday.

Chapter 9

"No great genius has ever existed without some touch of madness."

Aristotle

At the end of the summer we left Cleveland and returned to Buffalo. My departure was bittersweet in that I would miss those I loved, but not as much as one would think. I felt as though my heart was lined with Velcro and it was coming undone, unstuck and unglued. I didn't feel bonded to the people of the past—namely my father. I was enchanted with the people of the present, namely my stepfather. In letting my guard down with Tommy, I transferred my affections to the man with the steady paycheck, twinkling eyes and a house filled with books.

We ain't in Kansas anymore, Dorothy, and we damned sure ain't in Cleveland!

And we certainly aren't in love with a grown man who lost bits and pieces of his soul on the highway of life—which tragically stripped him of his ability to show up for his own children. By the time I had left Cleveland that summer, I was so close to being done with Charles Waiters. But it wasn't so easy to cast him aside because through my bloodstream ran an irrevocable surge of endearment for the man known as Charles Waiters. And this would cause great conflict within. In one breath, I didn't like him, but in the very next breath—I loved him dearly. And both breaths were at war. My emotions were a complex and steady stream of unprocessed and not fully understood thoughts.

Simultaneously, I struggled amidst an invisible war between me and my mother. And though no hard lines had been drawn in the sand or on top of the sidewalk—every time we interacted I felt as though I had crossed enemy lines.

"I want you to do as I tell you to do . . ." she insisted

"But we don't see things the same way . . . so that's hard for me," I defended.

"It doesn't matter how *you* see things . . . I'm your mother."

And? I wanted to say, but dare not, lest I lose a set of fresh teeth. It wasn't like I didn't love her. I did love her, and on the "sunny" days, I did all that I could to please both her and my father. Like every other kid in America, I wanted my parent's approval. *Approve of me and stamp me worthy so that I will remember for the rest of my life that I am valuable and irreplaceable.*

Somehow, this got lost in translation and I never received word.

"I'll make you proud that I'm your daughter," I wanted to bargain and reason with them. "But do me a favor and make me proud that you are my parents."

This too was lost in translation.

I never made them proud.

They never made me proud.

Instead, we existed on parallel streets that never had the common decency to intersect.

I could see them.

They could see me.

But for the life of everybody involved—we never seemed to find a way to stand on the same side of the street, the same side of an argument or love from the same side of our hearts. And this was our perpetual struggle.

So, it was a new day when I arrived back into Buffalo. I was externally the same but internally all together different. Upon my return, I was thrilled to see that Thomas and my mother had remodeled the house and bought us all new bedroom sets. The decorations in my room were so exquisite that I felt like Cinderella. This feeling elicited within me a measure of "good faith" that things were turning out for the better—despite my less than perfect relationship with my parents and my brothers, who were a non-stop annoyance in my life.

In the summer of '68 I earned a whopping $620 while working as an assistant for Uncle Willie. This was as close to being a millionaire that a ten-year-old could get, and I felt a profound sense of accomplishment by amassing such a fortune. Quiet as it was kept—I didn't tell a soul and hid the money in the floorboards beneath my closet. This was a little something I had lifted off one of those detective shows. "This should keep my money safe," I declared with a smile, basking in the glow of earning the money, saving the money and now hiding the money.

Once I got through with the business of being the business woman, I could return to life again as a normal kid. You know kid stuff—likes and dislikes, friendly competitions, rough housing and trash talking. In the neighborhood, I had a reputation for being the marble champion and would blow away the competition during our friendly little football games—which weren't so friendly at all.

My mother would stand in the doorway with her arms crossed as she watched under the weight of disappointment.

"Cheryl! Cheryl!" she'd call.

"Yeah . . ."

"Stop all of that rough housing and come inside."

"I want to play outside."

"Why don't you come inside and play with your paper dolls . . ."

"I'll play with the paper dolls later. I want to play football right now."

"Girls don't play football," she'd insist.

"This girl plays," I would say puffing out my chest with pride. "*And* she wins."

My mother would shake her head and call it quits.

Time to go in the house.

But I'd be back out tomorrow.

And she *knew* that—hence the bottomless pit known as her frustration—which would swell and come to life through the upturned corners of her beautiful lips, as they would take shape and form into what could fashionably be described as a frown.

My mother didn't care for my subtle or not-so-subtle deflection of gender roles. She didn't like it, but ultimately felt as though she had little control over it. I was destined to be who I was—whether it conformed to popular opinion or not. As a young girl, it seemed to me that adults always had this crazy notion about what girls were supposed to be. Their roles were over-defined in my opinion, with little wiggle-room for deviation.

I was an outcast AND an outsider for the simple reason that I didn't buy into the madness spewing forth from the genius masculine minds who had scripted feminine roles for societies' strict adherence. I did not belong to that world. I longed to be more free than crazy, and from where I stood, boys were free and girls were crazy for allowing a society ran by men to define them. Little boys were miniature men whose "home training" taught them that women were somehow a little less human than they themselves were. And little girls were coaxed into morphing into household servants as adult women—all in the name of being a good mother and wife.

Chalk it up to nonsense.

You can have it all!

None of it's for me.

It was during this time in my life that I met a neighbor across the street—whose name was Peaches. She was a simple girl who had a house full of brothers and was always locked inside doing chores and washing dishes. She could never come out and play—and only every now and again did she receive permission to sit on her front step and take in the world with wide eyes of longing.

This made such an impression on me.

Was life really this unfair for girl babies?

Peaches was little more than a maid servant to an entourage of men. And with every free step she could take just outside her front door—she moved in my direction. Ultimately, she became my best friend.

"One day . . . I'm going to move out of the neighborhood," she threatened.

Where you gonna go?" I'd ask.

"Anywhere but here."

And like most of the kids from our generation—we were all searching for a form of redemption and something to call our savior. Peaches looked for hers having a baby and I looked for mine in the books and when it was all said and done—our destinies would be much different. I continued to focus upon my studies and while attending School 39 in Buffalo, I excelled despite the fact that my brother was always in trouble, and it was me who always bailing him out. Ultimately, this would land me in trouble too.

However, I continued to court A's and B's while Chuckie preferred D's and F's. And yet still—I was "passed over" on praise while Chuckie and Michael bartered for all of my mother's attention. As usual, I lived on their leftovers—surviving on the "fumes" of what had not been completely evaporated by the neediest of my siblings.

Big sigh.

I was getting restless and in need of some stimulation. And right now, the Jackson 5 were in full swing and well on their way to becoming one of the greatest acts in the history of recorded music. They took the world by storm and when they made their national television debut—they blew a hole through our living room door *and* my heart and left in its place the "hope" of what is possible for people of color.

"Mama . . . I'm going to be a singer just like Michael," my brother Chuckie boasted as he hit a single note in the middle of the living room floor.

"Ooooh weeeee!" shouted my mother. "My son is going to make me rich!"

What? I almost wanted to ask out loud. The king of D's & F's is going to make you a mega fortune on a single off-note key.

Okey dokey.

I shrugged my shoulders and called it a wrap.

"What about me?"

"Don't you think I can make it big too?"

"What about me?"

"Can't I sing too?"

Sure.

Sure.

But I was not convinced. In that moment, I made up my mind. I decided that I would be the smartest person in the entire family (both living and dead) and that I would build my fortune upon the wealth of my brain. I would supersede all who had come before me and I would blaze a trail of genius for others to follow after I had gone. A noble goal for such a young girl—but I was serious about my intentions.

That's right.

I would show my mother in action and deed what my words were too inadequate to express. I'd show them all. I'd rise above my disappointment and become greater than the hole in my chest which longed so deeply to be filled with *somebody's* love. So, without further ado, it was time for another costume change—one that would be glorious enough to reflect my unleashed brilliance upon the world.

Brilliance became simple. It always is—and that's what makes it brilliant because most people overlook the uncomplicated situations in which we can tap our genius and call it forth. It came on the heels of my last year in elementary school in the form of a task requiring the "thinking mind" to render itself useful.

A single strand of rope strung from floor to ceiling contained a dollar bill at the top. A mass of students congregated at the base of the rope and struggled to remove the dollar. Each student was given one shot to snag the buck, but interestingly, they had all failed miserably. Caught between the desperation to obtain the dollar and the utter chaos of every botched attempt, I stood back and watched carefully every failed effort—all the while taking detailed mental notes of mistakes made during the process. Within no time, I realized that each student had approached the assignment in the same way—which spoke loudly to me that it was the "wrong way" because no one had been successful. So, "right way" or "wrong way" it didn't matter—I knew that when my turn came, I had to do it a *different* way if I were going to get the dollar. I noticed that all of the kids were grabbing low at the rope. Instead, I aimed high knocking ten feet off the climb.

I grabbed and held on for dear life, pulling myself up.

I got the dollar.

There was a spontaneous burst of applause as onlookers watched in amazement. I was cheered on and celebrated. I was also despised by those who were less than enthusiastic by the victory. To some students, me having the money meant that they didn't have it—and that didn't sit well with them at all. Others were just excited that somebody got it. The moment stood still in time as the feeling of being a little bit "smarter" than everybody else penetrated me. I could take this moment beyond the classroom—and I knew it. I could be smarter and I could get better results than most. That's what I learned from the rope and the dollar bill.

Be smarter.

Get better results.

From that moment on—I became obsessed with my own brilliance.

But this wasn't the only moment crystallized in my awareness for me to remember throughout the rest of my days. The onset of puberty brings great changes by wreaking havoc on your internal world as Mother Nature takes turns playing hop scotch with your emotions. I was up, down and all around on the emotional barometer.

Growing up can sometimes feel like a game of kickball—whereupon everybody's chasing after the very thing that's moving the fastest. And because of that—we are in a frenzy running on 'high emotion,' hoping that we might be the one who gets that "last kick in" before the game is over so that we can go home a winner.

I was doing that very thing when my life shifted again.

I was literally playing kickball with my friends, as I often did, when I got the call from beyond the stage that it was time for yet another costume change. However, this time I didn't initiate the transition, therefore, I had no idea that it was coming. Mother Nature did it for me and right in the middle of my greatest game of kickball I got my period and started bleeding. Simultaneously, church bells from across the street began to ring and the full-bodied harmony of resounding chimes echoed throughout the sky above me.

I stopped and stared at the bells.

I could not take my eyes off of them.

Nature called.

Then rang.

I heard.

It was mystical.

It was a whisper.

A gentle whisper.

Inaudible to the human ear but yet I heard it.

Not the bells—the call.

The bells were loud.

The call was silent.

You are woman now.

"Cheryl! Cheryl!" my friends called out to me, endeavoring to get my attention. I stood in a daze as I held the ball in my hands. The game hadn't ended, but I knew it was over. So I dropped the ball and without hesitation or explanation, I walked out of the game and into the rest of my life—never to return again. Instead, I moved on to others things like knock-down drag out PMS that left me with cramps so painful that the first few days of my period would render me helpless and bedridden. I even experienced 'black outs' and at times the pain became so excruciating that I would simply pass out in search of unconscious relief.

I moved on to other things too—like crushes and puppy love. I was in love with a little black boy named Raymond who lived around the corner. I didn't tell a soul about my feelings for Raymond. This was the season of innocent love and deep impressions were being etched in my conscious mind in observation of the external world. I saw that dark skinned men gravitated toward light skinned women. At ten years of age I was watching Jim Brown and Raquel Welch together on television and I'm getting the suspicious feeling' that white is right and light is just all right. I was light but my hair wasn't straight—and I noticed that there was a craze in the air for light skinned women with long straight hair.

I was a very attractive girl but looks never moved me like it did most women. I wasn't standing for hours on end in front of the mirror trying to make myself over as a beauty queen. Brains were my ticket out of an ordinary life. My mother and her sisters had the beauty queen role on lock down—and all of their "pretty" combined never took them anywhere exceptional. And at the end of the day, beauty was a down—right disappointment.

I sealed the deal on '68 with the saddest Christmas of my life spent in Cleveland. My grandfather on my father's side came back to celebrate Christmas with the family as a "bag of bones."

Literally.

His car stopped in the middle of Texas and he tried to cross the desert on foot. And somewhere between heaven and hell the earth opened up and swallowed him whole. Well, perhaps that's an exaggerated way to say that he died of dehydration and heart failure—but losing a loved one in such a "freakish" way can be a bit dramatic indeed. My grandmother took the bones and hosted a memorial in his honor with the bones on display.

What in the hell?

Hell?

Hell?

"You can't have a funeral with the bones!" protested the relatives.

"I don't see why not . . . it's all we got left of him!" grandmother cried.

Extended family members were upset and distraught—almost more about the funeral than they were about the death. It's interesting what pushes people to the point of no return. I just sat back and took it all in. It was like watching a ghetto Shakespearean drama with a poorly developed script and a lot of bad acting. If it weren't so tragic and unexpected, it might have been funny. Not what happened because "death" is rarely a humorous occasion but the *way* in which it happened could have been worthy of a small measure of laughter. Not a lot. Just a *little.*

The Lone Star state.

Car goes kaput.

Black man walking.

Black man down.

Bag of bones.

It's a wrap.

It was during this visit that Chuckie and I attempted to reconnect with our long lost daddy. Well, he wasn't really lost—but he was definitely long gone. But it still didn't fit well in my heart, so I wanted to try again. So, after our Christmas celebration, Chuckie and I cornered Charles Waiters with a request.

"Can we come and live with you?" we asked.

"No," he flat out answered.

"Why?" I demanded to know.

Of course, the explanation was shallow and void of great meaning *and* truth. And at the end of the day, it didn't matter the explanation, the answer was still no. My beating heart had taken enough and was bleeding to death from lack of love from my father. And though I can't say that he didn't love me at all—maybe on a "good day" he did love me a little—but he just didn't love me enough. Therefore, after a brief consultation with my brother, we both made an executive decision to forget about Charles Waiters.

Just forget him all together. That was easier said than done.

It always was.

Christmas was cut short that year when all hell broke loose on the streets of Cleveland. There was a shooting and the natives went berserk. The shooting proved to be racially motivated which only complicated matters and made them even more "ugly" than what they had already become.

It was during this era that the Black Panther movement was gaining popularity and membership. The tides were changing and it was Black power, Afros and "Picks" bearing the Peace sign. This was a lot to take in—and even as a little girl I was constantly approached and bombarded with the political messages of the day.

"Don't trust the White Man!"

"We need Black power, little Sister!"

"We need Black power!"

I was a kid with an afro, so they nicknamed me "Angela Davis" and assigned to me a philosophy that I could only pull off as a "costume change." I had the look but I didn't have the attitude, therefore, I remained a walking contradiction. I watched as The Black Panthers marched through the streets protesting, hell raising and tearing up our neighborhoods.

What's wrong with this picture? I was forced to ask myself witnessing the devastation with my own eyes.

You're burning down our world.

Both yours and mine.

The holidays came to a screeching halt. The riot was the interruption of everything and the aftermath changed me forever. The very next day, Chuckie and I were shipped back to Buffalo.

I returned to New York disappointed.

I also returned internally changed.

Modified and altered somehow.

I was still of African descent but I returned a little *less* Black.

Chapter 10

"To love oneself is the beginning of a life-long romance."

Oscar Wilde

Upon my return to Buffalo I began a new life but quietly.

"Cheryl . . . this is Miss Jeltz," my mother said to me, as she introduced a new character into my drama. This unassuming woman would enter the stage from the other side of destiny and the script would flip again. Jeltz was a member of the Seventh Day Adventist Church and she would come to our home and teach me and Chuckie about the Bible. Chuckie wasn't so keen on delving deeper into the Word, but I was rather taken by it all.

This was a grand departure from the customary sideshow at the Baptist church—which featured a cast of entertaining characters who performed regularly for the congregation. There was always the woman sitting dead center in the front row, wearing the oversized hat and blocking two thirds of everybody's view in the church. That was always her seat and she always wore that hat. And then there were those who flanked the side seats—loud, noisy and especially trained in the art of theatrics and dramatics. They were the ones who attended faithfully each week and *never* missed a Sunday service. They came armed with the three T's every time they stepped through the door: 1) testimonies 2) tissues and 3) tears. And then there was the "supporting cast." These were usually the people who partied all week and came to church on Sunday still in 'club attire' wearing fake diamonds and a lot of guilt. They also did a lot of crying and repenting on Sunday. They had redemption on deposit—and thank God they did because they went straight from the service and back to the streets pumped up on Jesus and Gin.

And then there were "the rollers." Every congregation had its designated rollers. They specialized in three very specific maneuvers:

a) The jump
b) The drop
c) The roll

And we can't forget the star of the show—the preacher. A highly animated figure adorned in holy clothing, vivid colors and sacred symbols. They walk with the authority of divine appointment and stand before the congregation carrying an expensive leather Bible that always looks ten notches better than anybody else's. But here's where it gets interesting—the preacher always starts the sermon by reading two or three versus from the Bible and from there, a two hour message is based upon their own opinion. But it doesn't matter, because usually by this time, the congregation is so high on the emotional intoxication of hysteria, no one seems to notice that nothing has been said; yet half of the members are down on the floor.

Crying.

Weeping.

Rolling.

And for what?

"What just happened here?" I always asked, in observation of the madness.

Millions of people have attended church for years and very few have ever read the Bible. So, how do they know what's in it? And aren't they the least bit curious? Perhaps everybody's too busy taking the preacher's word on it. These are just a few of the reasons that I was so drawn to Mrs. Jeltz. I wanted to adopt her ways because they were quiet and simple. They were brought to me without dramatics—I wasn't interested in being entertained, I was more desirous of learning. And by the time it was all said and done, I wanted to keep the Sabbath, reject pork and honor the dietary restrictions.

"Well Cheryl . . ." said my mother, "if you want a special kind of food . . . you're just gonna have to get a job."

"Okay," I responded without a quiver.

I was smart and didn't mind being a working girl. So, I got busy. At the age of eleven there weren't a lot of options for employment. But again, I was smart. I made my way to the local market and demonstrated my skills by retrieving the carts and stacking them back in order.

"I'm looking for work," I told the manager of the store.

He laughed.

"Said I'm looking for work," I repeated.

"You're too young kid," he said pushing past me.

But I didn't leave. I was determined to stay and show him what a big fat mistake he was making by passing me over.

"I said no," he said to me later that day.

"I heard you the first time."

"Then why are you still here?"

"Cause I need the work and you need the help."

Unimpressed, he pushed past me again till my mother came up to the market looking for me.

And once he laid eyes on Mary Strickland he was struck, as if knocked in the head by something big and blinding.

"That's your mother?" he asked, with stars dancing above his head.

"Yep."

"Wow . . . she's a looker," he said almost in a trance.

"Now what about that job?" I asked.

"Oh yeah . . . yeah . . ." he said, "be back here tomorrow morning and don't be late."

And just like that I was in.

I started work the next day stocking shelves under the radar. My employment was a quiet affair because I was under age. But that didn't bother me none . . . at long last I had enough money to buy my own food and keep the Sabbath. I was so sophisticated in my purchases that eventually my mother cleared a section in the refrigerator just for my food.

"Who's food is this?" Chuckie asked, with roaming eyes surveying the premises.

"Mine . . . and don't touch it!"

"Why you got your own beef?" questioned Chuckie.

"Cause . . . I'm special that's why," I boasted.

"Mama . . . Cheryl thinks she's special cause she's got her own beef!" he ran screaming through the halls. Well, perhaps now you understand that God inserted Chuckie into my life for comic relief. The only thing—it just wasn't so funny back then.

Chuckie and I were on very different paths and eventually we ended up at different schools. I was bused to an all white school when somebody had the bright idea that desegregation of the Negroes was a good idea. It began with School 49 and I would meet white kids who weren't fans of those with brown skin. Initially, I felt their dislike of me and their unfriendliness seemed to come from a deeper well within their being. And though they didn't know me, somehow their unkindness felt more personal.

But it wasn't all bad. There was a form of redemption amidst the madness. At School 49 I became an official member of the "Super Club." Everybody was referenced to as Super Something, depending upon what they excelled in. I was very intelligent and I also had a job, therefore, I became known as Super Ace.

Eventually I transferred from School 49 and was promoted to School 69.

School 69 had all white teachers and was in an all white neighborhood. This school made a profound contribution to my world in the way in which it molded me. I learned the rigors of a new math, higher learning and a different model of behavior. I was applauded because I was a global citizen with good grades, as I sucked the bottom dry in acceptance of all foreign philosophy. Slowly but surely I was losing the ways of the African. My new instruction was watered down with gentle subtlety to make it easier to swallow—as I became integrated into a world of white kings and queens.

Yes, slowly but ever so surely, I was becoming white.

Proper enunciation
Correct grammar
Etiquette
Manners
And respect.

This is what being "white" was all about, and I wanted it—to be all of those things and white. Alas, I was transforming into the respectable members of society that I watched on television. I watched *them.* I listened. I learned. I took their word on just about everything. And in return, they watched me back—but I had no idea till a few years later when I discovered that I was a "tracked student." A tracked student was little more than a guinea pig on the white man's chessboard. Ultimately, the system had great interest in learning whether Blacks were trainable. And with proper training, could they pull off being "successful" at this thing called life? I was filled with fury upon finding out that I had been tracked. There was no need to follow me around in the manner of a low profile Peeping Tom. If they wanted to know my destiny . . . all they had to do was ask.

School 69 offered much in the way of shaping my own personal destiny. I developed a close personal bond with a white teacher, Mrs. Russo. She had a profound effect on my life because she told me, "Cheryl . . . you can be anything you want."

And I believed her.

At School 69, I snagged my first boyfriend—Kevin. He was tall, cute and lived in the suburbs. He was a standout because he was the first "boy" that I actually wanted to claim as my own. I had crushes on boys before, but they didn't count. My first crush was a boy named Ralph when I was about eight and he definitely doesn't count because he was playing with paper dolls when I met him!

Scratch Ralph off the list.

Kevin could have been classified as the "Prima Madonna" of School 69. All of the girls wanted to be with him and all of the boys wanted to be him. But Kevin only had eyes for me—I was smart and he appreciated smarts. I was also cute, but I never paid attention to that. Like I said before, pretty never got Mary Strickland anywhere so I wasn't trying to follow that particular path. In fact, I fell off the beaten path during the Thanksgiving of '69 when a gang of relatives descended upon our Buffalo home to share the holiday with our family.

The men formed a single file line and headed straight to the pool hall. The women also formed a single file line but they ended up in the kitchen. The men's pre-Thanksgiving activities included pool playing, drinking, smoking and loud talking. The women's pre-Thanksgiving activities included cooking, cleaning, baking and slaving like dogs till sweat poured from their foreheads. When it was all said and done, the men returned from an afternoon of "good times," and sat down to a beautiful feast where they devoured the food like barbarians—leaving turkey bones, crumbs and dirty dishes as the hallmark that they had bothered to show up for dinner. The women were exhausted and left to clean. The men were full, happy and ungrateful.

They exited without so much as a thank you.

In that moment, I did declare that if this was what being a woman and a wife was all about—don't sign me up for the job. I wanted nothing to do with it! The point was further driven home during the Christmas of 69 when my mother received the following Christmas gifts: one lovely vacuum cleaner, one exquisite cooking oven and one delightful toaster oven. And upon opening all of her enviable gifts, she burst into tears.

"What's wrong?" my dear sweet stepfather asked with a puzzled look on his face. For he truly thought he was doing the "right" thing by her. After all, wasn't every modern-day housewife supposed to have the best of appliances to service the needs of her family?

"Am I a non-person?" Mary asked him.

"A non person?" he questioned. "I thought those things would make you happy."

Wrong answer.

More tears on the way.

Yep, I said to myself. *I definitely don't want the job.*

Marriage.

Kids.

A housewife.

And woman.

I won't check any of those boxes on my job application.

No thank you.

And though in retrospect, in looking back I can understand why Mary was so emotional. And being pregnant with her fourth child only added to her sensitivity. Yes, Mary was pregnant again. However, this time I was particularly fascinated with her pregnancy and enchanted with her changing form. Mother Nature had made her radiant, compensating for all of the inconveniences that such a condition might bring.

"I wonder if Tommy will love us the same when that baby gets here," Chuckie said to me one day out of the blue.

"I don't know," I responded.

Nobody knew.

We would all just have to wait and see. And we did—waited on pins and needles throughout my mother's pregnancy to see if Tommy would love his own offspring more than us. We pondered the possibility of being shuffled around emotionally and re-routed into the "stepchild" category.

In August of 1970, we welcomed yet another boy into our household. For heaven's sake, I couldn't catch a break. Following the birth of my baby brother, I began to feel the emotional oppression of not only being the oldest, but also the only girl. There seemed to be no relief granted to me because 1) I couldn't change my birth order and 2) I couldn't change my gender.

I felt as though I were serving a life sentence in a prison that had been constructed out of invisible walls. Nobody could see me—but I could see them from a distance as I stood within a world that had no windows, no doors, no ceilings or floors. It was an interesting room to find myself in—to say the least.

The offspring of Thomas and Mary, they named the youngest member of our family Keith. He was showered with every drop of attention, which meant that I got even *less* of the leftovers. But that didn't translate to being demoted as a stepchild. Thomas Strickland still loved me, Chuckie and Michael the same. And even if he did love his own kid more than us—he always seemed to have the good sense not to show it. Tommy was a stand-up kind-of-guy in that way and both Chuckie and I were relieved.

Once our family was complete, my mother wanted to seal the deal with a baptism.

"Cheryl," she said, "you're getting baptized tomorrow at church."

"Huh?"

"You heard me girl . . . baptized," Mary reiterated.

"Baptized!"

"Are you hard of hearing?" my mother asked.

"I don't want to get baptized!" I protested.

"Well that's too bad . . . you're getting baptized."

"I can't . . ." I said, chasing my mother down the hallway to explain.

"I'm not in the mood for no foolishness, Cheryl . . ."

"But mama . . . I can't!"

"Why?" she asked putting on the brakes so hard that I slammed into her backside.

"Because I don't want to catch the Holy Ghost," I said quietly.

"What?"

"Those people at church be catching the Holy Ghost and I don't want it!"

My mother took a good long look at me and let out a deep sigh.

Translation: end of conversation.

The next morning we were at church, and of course, I was getting baptized. There was no way out but through, so before they dunked me beneath the water I grabbed my nose and covered my face to "keep that ghost away." The congregants were baffled by this display and my mother was not pleased at all. In fact, she was downright furious. Needless to say, the Holy Ghost was a no-show and I went home with a big cheese grin on my face.

Sorry mama!

My baptism represented the end of one era and the beginning of another.

The 70's made deep impressions that left lasting marks that I'm still trying to erase. This decade paid homage to women and they were everywhere and represented in mass on television in roles of independence. Shows such as *That Girl* and *The Mary Tyler Moore Show* celebrated women's achievements. Women were clever enough to be detectives on television and get paid for doing it. Shows such as *Get Christy Love, Police Woman, Charlie's Angels* and *Wonder Woman* set the stage for women having brains and beauty. The 70's unveiled a new generation of Black faces as Hollywood rolled out a whole new crew such as *Julia, The Jeffersons, Good Times, What's Happening* and *Sanford and Son.* However, even as a child I didn't bond with these shows because I could feel they somehow glorified the plight of "struggle" within the African American community.

Julia was a single mother trying to raise a kid by herself.

I didn't like that.

Good Times was a show about "bad times" in the ghetto.

I didn't like that.

What's Happening should have been titled *"What in the Hell Just Happened?"* because that's what I felt like every time it came on television.

I didn't like that.

Sanford was an uncouth businessman man in Watts and his *Son* was a stone's throw away from being an idiot.

I didn't like that.

And although *The Jefferson's* were moving up and they had a cleaning lady to show for it—they still never seemed to get anywhere.

And I didn't like that either.

Hollywood did little to help promote the image of positive black role models. Instead, we were depicted and wildly celebrated as pimps, players, thugs and hoodlums. Our music co-signed this image with hits such as *Superfly* popularizing the notion that a Black men selling drugs and pimping women could now be categorized as a legitimate business pursuit.

And I damn sure didn't like that.

Black women were less threatening to White America, therefore, they were given more opportunities. Black men were silently demoted as the breadwinners of the family, and in earnest attempt to recover their position on the throne—many took up the business of dealing drugs. Ironically, the influx of drugs into the inner cities gave our Black men a sweeping opportunity to become entrepreneurs. And sadly, a lot of them took it.

During this time, I fell under the spell of Billie Jean King, a professional tennis player who challenged, battled and championed gender discrimination. As a pre-teen I followed her matches—on and off the court in the quest for gender equality. I felt as though I did the same thing every day in my own house that Billie Jean was doing in front of America. The only difference was—she was on television and I was not.

She was getting paid and I was fighting for free. But my struggle was not in vain. I was on the crest of a new discovery in understanding myself and the world around me. I turned to images of movie stars in an effort to see myself better. The Jackson 5 was blowing up the world and I saw literal "magic" in what they were doing. For a beat of time, I too, contemplated the journey of an entertainer. So I gave it a shot by joining an all girl band and we hit the talent show circuit. My career as a singer was short-lived but I did get the opportunity to do some background vocals for a guy who hailed from Buffalo. At the time, he was a local wanna-be-superstar who had his own brand of funk/soul. Ultimately, he went on to become a national sensation. His name was Rick James.

But like I said before, my singing career was short lived. Just before my 13th birthday, I received a Christmas gift from my mother that would change the ground I stand upon. I got a microscope set. There was something about this contraption that was both mystical and utterly fascinating to me. I was enchanted in learning about life beyond what the eye could see.

From that moment on, I declared myself to be a young scientist who would go on to become a mechanical engineer. "I'm going to build a machine that will change the world," I declared.

But first . . . I had to change *me.* It was during this time that I was stopped on the streets by tambourine toting Harry Krishna's wearing peach robes and bald heads—who began a dialogue with me that I would continue in my head long after they had gone.

They had posed questions I couldn't answer.

They made me think.

They made me ponder.

What does it mean to love myself?

It was time for me to learn.

Chapter 11

"In dreams, we enter a world that's entirely our own."

Harry Potter

In the summer of 1971 I graduated from middle school. The event itself was less memorable than the significance of what was going on all around me. This was a transitional moment in my life and much like a bridge—it would serve to connect this world to the one that lay just on the other side of it. A course of destiny was set into motion and during this time, my father did something unprecedented. He drove from Cleveland to Buffalo and attended my junior high school graduation. His presence was surreal in more ways than I can recall. And though he was polite and brought pleasant conversation as his most honorable gift, there seemed to be much said beneath the surface that was inaudible. However, I will never forget the words that rang loud and clear as Charles Waiters sat me down to offer a bit of "fatherly advice" on my journey's way.

Listen Cheryl . . . if you want a Cadillac, don't talk to the man with the 1957 Chevy.

Anything you want to know you can learn it at the library.

Grab the tiger by the tail and don't let go.

Be number one and save your money.

This was the extent of my father's contribution to my life—and yet somehow, I never forgot it. Though later in life I would learn that there were greater lessons that he either forgot to teach us or he simply didn't know. Despite it all, I still loved my father very much and so did Chuckie. And no matter how hard we tried, we just couldn't get away from being *in* love with someone who just couldn't love us back the same.

Ever.

Ever.

Ever.

"We want to go back to Cleveland with you," I said to my father before he left.

"I can't do that," he said. "You belong here with your mother."

"Don't we belong with you too?" Chuckie spoke up to ask.

"Of course you do," he said trying to ease our disappointment with a polished smile that looked to be more rehearsed than anything else. "But you belong *more* to your mother."

His words spilled, rolled and dropped onto the floor harder than they had ever done so before.

Chuckie and I were done.

Truly done this time.

And when Charles Waiters left Buffalo, Chuckie and I stood on our step and watched him drive off.

"I don't think he cares about us at all," Chuckie said.

"Maybe you're right," I said readily agreeing with him.

We were beginning to sound like broken records, and it was time for the scorned children of yesterday to move on with their lives.

I knew it and Chuckie knew it too.

"When we grow up we're gonna be rich, Cheryl . . ." said Chuckie, "and then we're going to buy Charles Waiters' street and put him out of his own block!"

I laughed.

It was funny.

We just might.

The thirteenth year of my life brought significant and long lasting changes to the person I would eventually become. In the midst of exploring who I was—I also began to redefine who I believed I could be. In some ways, it could have been said that I put the "cart before the horse" and growing up seemed inevitable now. In the midst of my ordinary childhood something extraordinary happened.

I went to Yale University!

It wasn't an accidental visit nor did I land upon the prestigious University's grounds in New Haven, Connecticut as a figment of my own imagination. It was a real manifestation of a dream that I grabbed hold of and dare not let go. As an exemplary student in the Minority Mechanical Engineering program, I was afforded the opportunity to spend the summer of '71 at Yale in order to pursue my passion of mechanical engineering.

Mechanical Engineering afforded me the opportunity to flirt with electricity, machine design, math and science all at the same time! I had found my own version of Utopia. The moment I stepped onto the immaculate grounds I knew that my world was brand new. The college was massive and so was my ever expanding world.

"I can be something great!" I told myself. And from this space I began to carve out my future—blazing, bright and glorious as ever. I chartered my course with a precision which included my dream career of mechanical engineering, a husband, a quaint home with a white picket fence and 2.1 children—though I never quite figured out how to get the 2.1 kids.

I felt the spirit of invincibility and believed that there was nothing I couldn't do. I was making an early withdrawal on my brilliance and it would take me all the way to the top. At thirteen years of age, I was certain of this. Though later in life I would come face-to-face with the sober recognition that invincibility may have been a close kin to disillusionment.

During my tenure at Yale University, I helped a team of engineers construct a prototype for offshore drilling wells. This would set the precedent of more "extraordinary" moments to come—which included a second internship spent at the Lehigh University in Bethlehem, Pennsylvania in the summer of 1972. Both opportunities put me on the fast track of growth, learning, maturity and sent me far ahead of my years. But would I remain there? Only time would tell.

Once the euphoria of Yale began to wear off, I came crashing down and was looking for a soft pillow to make a hard landing.

"I have no money to send you to a private high school," declared my mother.

"What?" I cried out in a panic.

"With all of the boys . . . there's just not enough money."

I was devastated. Throughout my middle school years I had always assumed that I would go to a private school. I had always excelled and demonstrated exceptional talent in the world of Academics, so how could this possibly be? There wasn't a lot of time for debate because I had to raise some money for a quality education so I got on the phone and put an emergency call into my father.

"Daddy . . . I need some money for a private high school," I began but I was quickly interrupted with a pile of excuses which added up to little more than a long run on sentence.

Don't have it.

Me and my wife split up.

Business has been slow.

Running short on just about everything.

Click.

I don't even remember saying goodbye.

When it rains, it pours. It also crumbles and falls completely apart, but most people are too busy walking around in a self-induced coma to ever bother with the details of a life that is slowly undoing itself at the seams. In short, this is what was happening in our home.

From the moment Charles Waiters left my graduation it was though he left some voodoo curse that sat right in the middle of our living room floor waiting to take possession of an unwitting soul. My mother was oddly *different* upon his departure. My stepfather was different too, but nobody talked about it and while standing in the same room, they began to separate quietly—as if being pulled apart by something that neither of them could see and with no way to bring themselves back together again. They were growing apart and the house became noisy with the deafening echo of their distance from one another. And when Thomas stopped watching Gunsmoke with the family, I knew all hell had broken loose—but it was a very quiet chaos.

I wasn't too concerned because I had affairs of my own to tend to—namely, my first real love, Dwight Brown. I met Dwight when he came to my house to pick up my brother for a baseball game. I was sitting on the porch and minding my own business when this young man pulled up to our front door driving this tan Electra. And when I laid eyes on him, my heart stopped at the sight of him. A light skinned version of Burt Reynolds, Dwight Brown had a cool that was too smooth to deny.

"Who you looking for?" I asked him all the while trying to be sexy wearing my favorite pair of shorts.

"I'm here to take Chuckie to the baseball game," he quickly responded.

"I'll get him," I promised with a little tease in my step.

"Okay," he said slowly, feeling the vibe of me digging him and returning it in full measure.

He's gonna be my boyfriend, I said to myself as I went and got Chuckie, but not before inviting myself to the baseball game.

"What?" protested Chuckie, "You can't come!"

"Shut up and get in the car!" I said offering him a 'love tap' on the back of the head.

I rode to the game in Dwight's car and by the end of first inning—I was applying for the job of his 'girlfriend' by showing him how good I could kiss. Dwight was accepting applications on the unfilled position of girlfriend, so he didn't mind me showing him my skills. By the time he dropped me back at home, I had got the job. Needless to say, this was privileged information, so nobody knew that I had a boyfriend.

Dwight was a seventeen year old charmer who teetered the line between man and boy, though he was closer to a man than a boy. He also used his more experienced and worldly ways to introduce me to the grown up world of sexuality. In turn, I lost my virginity at the age of fourteen and began sneaking Dwight into my bedroom at night. It was a big risk but I was in love and was willing to take the risk. No one ever knew what went on behind the closed doors of my childhood bedroom. But I can tell you this—I wasn't a child anymore.

Women's Lib was in full swing by the time Dwight and I got together and I caught the fever of liberation and didn't mind boasting my ripened figure by trading in my Tomboy ways and unflattering skirts for hip huggers, sassy shirts and boy sneakers. I was bringing sexy back before it ever went out of style and I liked it like that!

I was growing into being my own woman, but this wasn't the only thing that was developing. My best friend, Peaches, from across the street had a baby growing in her belly and when my parents caught wind of this on the heels of discovering that I, too, had a boyfriend, they sat me down for a come-to-Jesus-kind-of-talking-to.

"We want to talk to you about boys . . ." my mother started by leading the conversation.

Thomas didn't offer much by way of words—only support by throwing out a nod or two every now and again.

It was safer that way.

"What about boys?" I asked.

"We know that you have a boyfriend and we don't think you should be seeing that boy!" my mother came right out and said.

Thomas offered a single head nod.

"You're just too young!" injected Mary.

"Too late . . . we're already in love!"

Their eyes widened like flying saucers and my hard stance and inflexible position cut them down half-to-size. With an attitude like that, Mary and Thomas were smart enough to know that they weren't going to get anywhere with me. I had my mind made up—I was in love with my boyfriend and pissed off at my best friend.

"What did you go and do that for?" I questioned Peaches when I finally caught up with her.

"Up and do what?"

"Get pregnant?"

She shrugged her shoulders and just like that she gave up on her dreams. Once upon a time ago she dreamed of being a hairstylist, but now the air had been let out of all that she had hoped upon. She went from being a full time slave to her family to being a full-time slave to a newborn. What little joy she had left was sapped by a crying baby, and just like that, she fell out of my life and into motherhood. But as for me, it was a blessing because I was able to get a behind-the-scenes peek at the staggering reality of teenage pregnancy and it definitely wasn't for me. I had other plans. Dwight and I were going to build a life together. I was going to be an engineer and he was going to be a pediatrician.

In the meantime, I began high school and immersed myself in the culture of the debate team, choir, chess, swimming and theater. I consistently made honor roll and began to set my sights on attending M.I.T. following graduation. But there was always an issue surrounding finances so I knew I had to figure out a way to amass a fortune and get it done. Just when I thought all hope was lost, my Algebra teacher hooked me up with what I thought was a great opportunity till I hit the door.

"Cheryl . . . how about a babysitting job?"

"Oh yeah . . ." I agreed, "babysitting is fun! I'd love to do it!"

"Great," the teacher said. "You'd be babysitting for one of the players of the Buffalo Bills."

Even better, I thought to myself—till I showed up at his door and discovered that he was short-tempered, impatient and not particularly kind to his wife. His name was O.J. Simpson and his family was living in an upscale suite in the Jamestown Apartments. My assignment with the Simpsons was short lived and I knew that it wasn't a viable way to secure my tuition into M.I.T., therefore, I had to find alternate ways.

Perhaps I could acquire superstardom or an alternative source of funding as a television star, so I appeared on a television show called *Academic Challenge.*

My television debut was nothing short of disastrous. I was blinded by the bright stage lights and when the host asked me a question I literally froze.

Froze.

And froze.

I never thawed out and concluded the show much in the fashion of a zombie. I was horribly embarrassed and cried for two days straight. It was a harsh realization that neither superstardom nor television were my calling. Well, I still got mathematics, microscopes and the dream of M.I.T.

Okay.

Moving on.

By the time I reached the age of fifteen, my life had begun to unravel all over again.

Hold on mama. Why do I get the feeling we're about to land on the steps of an old bus station at 2 a.m. in the morning?

My mother was growing cold and more distant with the passing of each day. I would find her staring oddly out of the window—a million miles away from earth or so it would seem.

"What's wrong with you?" I asked.

"You ruined my life," she mumbled to me one day.

Her words ran through me like the pristine blade of a Roman soldier's sword.

And what was left of her and I died upon that day and that death was imminent in all that perished around us.

Caesar, the family dog ran away from home never to return.

Mary and Thomas appeared more as vapor than real people.

The Vietnam War was drawing to a close but the one inside of our home was just beginning. Upon hearing word that Thomas's stepmother had died, he made plans to leave town in order to attend the funeral. The moment his car turned around the block, Mary backed a moving truck into the driveway and began to move us out of the house.

"What are you doing?" I screamed at her.

"We're moving!"

"I don't want to move!"

"Move it, Cheryl!" she said, as I parked my behind on a chair and refused to move it.

"Why are you doing this?" I demanded to know.

But there was no answer.

There never was.

I knew she wanted to be free. She didn't want to be married anymore. She was hitting the clubs on a frequent basis and had already demoted herself from parent to buddy.

"I thought you loved him!" I shouted. "I thought you loved him!"

Silence.

"I don't want to move again!"

Nothing.

I held my position till me and the chair were the only two items left in the house. And frankly, Mary probably debated on whether or not she could take the chair and leave me behind.

But I surrendered and followed Mary and her young, punk, no-nothing boyfriend across town to a raggedy apartment where we sat up shop and played house all over again with a broke man who had even less than Mary.

Here we go again.

I choked on the thought of my mother and her unattractive ways. She mated with her men and then she "killed" the relationship. I began to address her as the black widow. It didn't go over well, but I didn't care. I couldn't help but wonder if she ever loved Tommy or had she just been pretending all of these years—so that he could be an agreeable meal ticket without much resistance for the fact that she had little to no heart at all.

When Tommy arrived back into town he flipped a lid when he discovered that Mary had moved out. He tracked us down like a bloodhound and came calling for what was rightly his. But all that he could claim was my baby brother, Keith.

"I want my son!" he demanded.

"Get off my step Thomas Strickland!" my mother shouted.

"Give me my son!"

My mother wasn't willing to give up Keith, but that didn't stop Thomas—so he took him by force. Barged right in, picked him up and left.

You bring me back my Damn son!

You bring him back here!

There was so much commotion that Mary ran back into the house to call the police. In the meantime, I jumped out of my bedroom window and ran into the middle of the street and threw up my hands and started screaming, "Daddy! Daddy! Daddy!"

Thomas stopped his car on a dime, backed up and opened the door . . . but just as I turned to look over my shoulder there was Mary Waiters pushing up the street on the edge of what looked like a runner's sprint! Breathless, I quickly grabbed the handle, opened the door and jumped inside with barely a moment to spare! As I slammed the door shut and "locked it" Mary had reached the car—and was pulling on the handle with all of her might.

"Drive!" I screamed.

"Drive!"

And with that we sped off leaving Mary in the dust running after the car exhausted and exasperated. And I can't deny the truth—I was happy. I was happy to see me going and even happier to see her staying behind.

Bye bye Mary Strickland.

Bye bye.

I moved in with Thomas Strickland but my happiness was short lived. Life had always seemed to work out like that for me. The court ordered that Thomas return my baby brother to Mary and he did. I was granted permission to stay and I lived with Thomas for one year.

At sixteen years of age I was emancipated from Mary Strickland and Charles Waiters by the state of New York and given my child support payments of $18 per week.

I got a job at H. Salt Fish & Chips and disowned Mary.

Mary disowned Thomas.

And at least for a period of time it would appear that all of life went to hell in a hand basket.

north

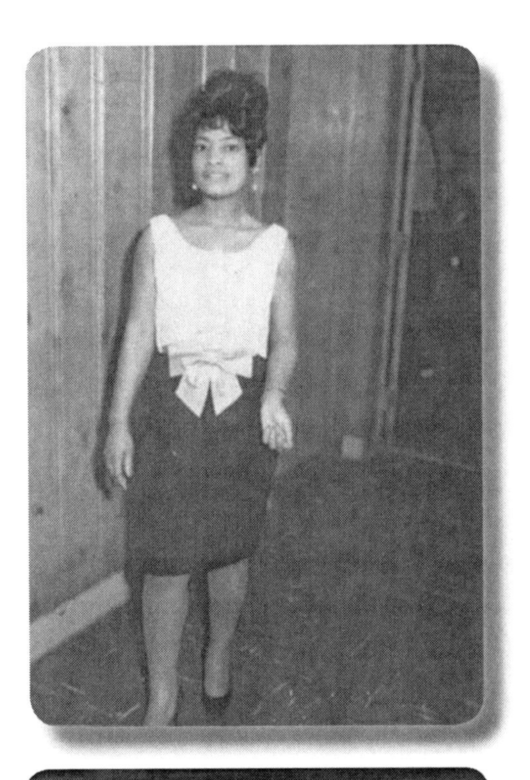

Chapter 12

"This was definitely not my destiny."

Cheryl Waiters

When the sky fell down, I tipped over into the arms of my boyfriend, Dwight, and I was safe there. But security, much like everything else in my world was temporary at best.

"Your boyfriend is pimping ho's," I was told by Dwight's best friend, Jeffrey, on a day when I was the least prepared to get knocked off my feet by a runaway fly ball.

"My boyfriend is what?" I asked as my jaw dropped to the floor.

"You heard me . . . pimping ho's."

"I don't believe you!" I scoffed. "You always been jealous of Dwight."

"Then why don't you go down to the corner in front of the liquor store and see for yourself."

"I ain't even going to waste my time taking the walk," I boasted indignantly.

"Cause you scared . . ." he taunted.

"I ain't scared of nothing . . . least of all that foolishness you talking."

"Then take a walk," he insisted putting it back on me.

I pressed my gaze against his and found that his eyes had no wiggle room.

Interesting, I thought. *Could Jeffrey really be telling the truth?*

So I took a walk—down, round and just beyond the corner. At first glance, there was nothing. At second glance, still nothing

That chump Jeffrey, I said to myself as I turned to walk away and there out of the distant gaze of my left eye, I saw Dwight on the opposite corner—who looked to be directing traffic with a handful of working women flanked by his side.

"What the????"

I couldn't get across the street fast enough my eyes burned a whole straight through him as I watched every turn of his cheek and crease of his smile as he took and gave, passed and held dollars in his pocket. It seemed as though everything on this corner was up for negotiation.

Well, everything except for *me.*

"Dwight! Dwight!"

When he saw me, his eyes widened and expression froze somewhere between horror and shock.

"What are you doing here?" I demanded.

"What are *you* doing here?" he asked, meeting a question with a question.

"Jeffrey told me you was down here pimpin' ho's!"

He tried to put his arm around me to direct me away from traffic. It was probably bad for business to have the pimp's irate girlfriend calling him out on a public block.

"Baby . . ."

"Don't baby me! What are you doing down here?"

"Just making a little money. This is a side hustle."

"Side hustle?" I blurted. "What about our plans?"

"This ain't got nothing to do with that."

"You said you wanted to be a pediatrician!" I shouted.

"I know . . ."

"Well this ain't got nothing to do with being no doctor!"

"I'm just making some dollars, baby . . . it ain't no big deal."

His words cut through me like wind laced with fire, and I knew in that moment it was over. I could hear Dwight talking but his empty excuses bled into the noiseless sound of my feet as I turned in the opposite direction and crossed the street alone. This was not the man that I knew and if it was—he and I had little to nothing in common.

I wanted more for myself.

I had dreams and aspirations.

Dwight had hustle.

I had goals and lofty ambitions.

Dwight had hustle.

We were *uneven* at best.

"Baby! Baby!" he shouted.

"We don't want the same things . . ." I mumbled.

"Baby! Baby!"

"We don't want the same things . . ." I repeated as tears followed every step I took all the way home.

Back at home with my stepfather things were becoming as shaky behind closed doors as they were on the street. Thomas had started drinking and the alcohol changed him almost overnight—for the worst. Initially, he began with propositions that felt out of line and inappropriate.

"Cheryl," he began by saying, "You can have everything your mother threw away."

"What???" I gulped.

"You can have it all."

"I don't want any of those things," I told him, "I just want to go to college."

"I'll give you whatever you want, baby . . ." he reassured me in the beginning, however, his reassurance quickly turned into what I referred to as a frightening misunderstanding. In drunken fits, he began to confuse me with my mother.

"Mary . . ." he called out to me. "Mary!"

This is no good, I said to myself. *I've got to get out of here . . . at least for a while.*

So I landed on the doorstep of my schoolmate Alisha Davis. Alisha came from a large family and they were gracious enough to offer me a place to lay down my head at night.

"You can stay with us for a while, Cheryl," Alisha promised.

"Of course she can," her mother agreed.

I felt a sense of relief and graciously accepted their invitation to move in. I thought I had done the best thing for myself until I realized that Mr. Davis was making eyes at me from across the dinner table. His long and overly-friendly stares had become downright creepy, and when I walked through the house I could feel him gawking at my backside. And though I tried to make sure that I was amply covered, I couldn't hide the fact that I was fitting more into a woman's body than a child's.

In looking ahead, I could only see potential disaster in staying with the Davis family any longer. I had to do something and I had to do it quickly. Out of desperation, I caught the next thing smoking out of town and headed back to Cleveland. However, this time I was on my own.

And sadly, my departure was grossly misunderstood by those I left behind as well as those that I returned to. Everybody said that I ran away—but I didn't run away.

I left.

My arrival into Cleveland was an unwelcomed event.

"What are you doing here, Cheryl?" my grandmother questioned.

"I had to get out of Buffalo," I tried to explain to her and Uncle Willie.

"But you can't just up and leave your whole life," Uncle Willie said to me.

"I didn't have no life there!" I spat.

"You had plenty of life," insisted my grandmother, "and you ain't had no good reason to leave it behind!"

It was this kind of thinking that caused them to ship me right back to Buffalo, and there I was again on my stepfather's doorstep—praying on a miracle or restitution, whichever came first. My mother was completely out of the picture and for all intense purposes Tommy seemed to be the only family I had left. And though his behavior was erratic and unpredictable at times, I managed to make it to the summer of 1975 whereupon I would return to Cleveland again. I was seventeen years old and there was no way in hell that I was going back to Buffalo *this* time—which only infuriated my father.

"You can't stay here in Cleveland!" he shouted.

"Why not?"

"Cause you belong at home with your mother!"

"I haven't been with my mother in a hundred years!" I retorted, "And if you ever bothered to call you would *know* that!"

"Don't you be getting sassy with me, Cheryl!" he warned. "You just can't stay here in Cleveland. Ain't nobody here to look after you!"

"What about you?" I pressed him, "you're my daddy! Ain't you?"

"I got a lot going on," he rambled with another long and drawn out, old and tired story of why he couldn't do this and couldn't do that.

"Story of my life," I mumbled. "I ain't going back to Buffalo, daddy. I'm here on a mission. I'm planning to go to college . . . so we're just going to have to deal with me being here the best we can."

The firmness of my voice let everyone in the Waiters family know that there was no room for negotiation.

None.

"Okay," said daddy reluctantly, "You're going to need a job!"

Alas! I was thrilled to finally hear that Charles Waiters had come to his good senses. At the end of the day, his good senses included getting me a job at McDonald's and room and board at Mr. Camp's house. By the way, Mr. Camp was a seventy-year-old widow who lived two doors down from my grandmother. He was an extra creepy old man who had an affinity for walking through the house in the late night hours, knocking on my door and peeking in on me. His house smelled of "old age" and arthritis, and I spent the better part of three months in his home trying to figure out how to get the hell out. This little arrangement was my father's answer to everything. If this was the best he could do, I knew that I had to jump and intervene on my own behalf—and that's exactly what I did—intervened—with a little help from my Muslim friend along the way.

"Cheryl . . . you wanna come and work for me?"

It was an offer I couldn't refuse. It would get me away from flipping burgers and earn me a decent enough wage to rent a studio apartment where the bed was built into the wall and I could pull it down to the floor. It was like heaven—my own piece of humble real estate.

Alas, I had found my own place in the world, and nobody was going to take it from me without my consent.

In the fall of 1975, I enrolled myself into a local high school to finish out my senior year. I was back on track and it felt good to be in command of my own destiny again. For so long it had felt just out of reach, but now I could feel the winds of change sailing in my favor once again. Shortly after the semester began, I met a student that I was strangely attracted to. His name was Garey Berry and he was odd in a peculiar sort of way. A first cousin of the famous actress, Halle Berry-Garey referred to himself as "Ziggy." He was a handsome light-skinned man standing about 6'1 in height. An intelligent hippie, Ziggy was a rebel and a free spirit both at the same time. He had an exquisite compassion that called me to him—though I never quite understood the essence of the call. Throughout the course of our relationship, we developed an unusual bond. I called it "unusual" because I was enchanted by him on a spiritual level. I desired his friendship and longed for his companionship when we were apart. He was always welcome in my world, however, there was another side of him that I struggled to engage: the physical part.

I just wasn't physically attracted to him, however, the lines between the physical and spiritual were so blurry that it was hard to pinpoint exactly where my "liking" of him as a friend and my "dislike" of him as a lover began and ended. It wasn't hard for Ziggy though, and he knew exactly what he wanted.

"I want to marry you," he said to me one day.

"Really?" I asked with a look of shock plastered across my face.

"Yeah . . ." he continued, "and I want us to have kids."

"Oh," I replied without a lot of emotion. "Let me ask my daddy."

"Daddy . . . can I marry Ziggy?"

"Whose Ziggy?" he asked with one eye cocked open wide.

"A boy from school."

"He got any money?" Charles wanted to know.

"I don't think so."

"Then hell naw you can't get married!" he stated flatly.

"Okay," I said shrugging my shoulders.

Guess it wasn't my destiny, which didn't bother me one little bit. Ziggy was stuck on genetic breeding and wanted to hand pick his biological partner to produce a certain kind of offspring. He was interested in beautiful/smart babies and he thought that I fit the bill on his high dollar biological experiment.

This was definitely not my destiny.

Ziggy up and disappeared one day and I never heard from him again. I never knew what happened and just presumed that his departure was a cowardly way of saying goodbye. He had left me in the dust on the heels of all the words he could not say—at least this was my way of reasoning his vanishing without an explanation.

Later I would discover that Ziggy had joined the Air Force. I found this out quite by accident one day while cleaning off my father's bar, and

I ran across a slew of old letters that had been sent with money. But the money was spent and letters had been hidden by my father.

Charles Waiters!

How could you be so shady?

I was thoroughly disgusted and between Charles and Mary, I was ready to divorce them both and take out an ad in Sunday's paper for a new set of parents. The job description would have read a little something like this:

Brilliant young woman interested in bettering her life seeks two mentally competent and fiscally responsible adults to apply for the position of parents. On the job skills include compassion, a sense of responsibility and a willingness to show up and participate in my life without a host of personal dysfunctions (i.e., drama) and hidden agendas and itineraries. Co-dependent women and hustlers need not apply. No experience necessary. Position to start immediately.

Charles Waiters never offered an acceptable explanation as to what happened to the money and the letters that had been sent by Ziggy. In fact, he was so caught up with his new girlfriend, Janice that he didn't have time to pay too much attention to my tantrums. Infatuated again, Charles lost all footing in the real world when he went "koo koo for cocoa puffs" over his new love. Everything was colored to the tune of Janice, who was an interesting woman with a lot to say. In fact, she took it upon herself to teach me the four principles that every woman should know. She taught me how to a) look young b) act like a lady c) think like a man and d) work like a dog.

I was a willing student eager to take on the challenges of womanhood and found her curriculum interesting, but at the same time I was getting a more risqué education from my cousin Rhonda, Uncle Willie's daughter. She worked at the local gas station and had an affinity for older men. She was dating an old man who owned a record store and between the gas station and the record store, Rhonda was making a killing. She was literally robbing both places blind!!!

"Oh Lord . . . this is bad on my nerves!" I told her one night while I was hanging out with her during one of her robbery sprees.

"What you whining about, Cheryl? This is how you make that paper!" she boasted, stuffing money in her bras.

"What if you get busted?" I asked her in a panic.

"Relax Cheryl . . . I ain't gonna get busted. This is what I do . . . I'm a professional."

"You're a professional thief?"

"I'm a professional working woman," she declared.

Needless to say, that relationship didn't last for long. Most of my relationships didn't and people just seemed to drift in and out of my life—or maybe it was me who was drifting in and out of theirs.

I went on about the business of living till one day my life changed. The Stop 'n Shop where I worked was robbed at gunpoint, and as I stared down the barrel of a 38 caliber pistol—the whole world took on new meaning.

"Give me all of the money, Bitch!" demanded the robber.

Terrified, I quickly opened my drawer and surrendered everything down to the last rusty nickel. And if the hooded robber had time, I would have handed over all the pennies too.

When the robber left, I immediately called the police and the owner—who was furious about me handing over his earnings to a masked gunman.

"Excuse me . . ." I said, "But he had a real gun. It wasn't a water gun. It was a REAL gun!"

"Well how 'bout this . . ." he said, "You're REAL fired."

"What?"

"That's right. Quitting time."

And just like that I lost my job, my income and my home. I was about to be out on the cold streets till I met Carmen a woman in her early 30's who was working in an ice cream store. She was a Muslim woman and a teacher who had taken a gig scooping ice cream in between teaching assignments.

"What's your name?" she asked.

"Cheryl," I responded one day while going in for a scoop of ice cream.

"You live 'round here?" she asked with a smile.

"Not for long," I said with a sigh. "Just lost my job and I'm about to lose my place."

She looked at me for quite a while before responding.

"Where you gonna go?"

"I don't know," I said looking despondent.

"How 'bout you come with me?" she suggested.

"Huh?"

"I got a place you can stay at for a while," she offered.

"Really?"

"Yeah . . . it's a decent place. You'll like it."

I paused and contemplated my options. During the contemplation, I realized that I had no options.

"Okay," I surrendered.

A decent place beats the street any day of the week, so I was sold on it—at least until I got there and the script flipped introducing me to a whole new game. It was something that I was ill-equipped and unprepared to deal with on many levels.

On any level.

Shortly after my arrival Carmen began to act strange. When I came home in the evening, she would have all the lights cut out in the house and candles burning from the bedroom to the living room.

"Carmen!" I called out to her. "Carmen!"

For a moment there would be nothing, not even a shadow and then she would appear—almost like a black cat toying with midnight magic and voodoo spells. Mysterious and mischievous, she would slink her way into the living and prance around half-naked.

What in the hell is going on around here? I asked myself, terrified to make a move or a sound.

"Don't you like me?" Carmen asked, appearing in the middle of the living room wearing very little clothing.

"Uh . . ."

"I like you . . ." she began seductively.

"Sure?" I replied more as a question than an answer.

"You like me too, right?"

"I don't know what you mean by like . . ."

She grabbed my hand and led me into her bedroom where she was insistent on showing me her version of *like.*

"I don't know what's going on here . . ." I stuttered.

"Relax . . ." she insisted. "A woman can offer you a better relationship than a man."

"I ain't never thought about a woman like that . . ."

"Well . . ." she said with sass in her voice, "a woman offered you a place to lay your head down at night, right?"

"Yeah," I sputtered slowly.

"Not a man," she concluded.

"Yeah," I repeated, not sure where she was leading with the topic.

"And that beats being homeless, right?" she asked.

"I guess."

"You guess?" she pressed. "It's best to k*now* these kinds of things, Cheryl . . . if you want to keep a roof over your head."

And just like that, this was the beginning of something foreign and downright uncomfortable.

Dorothy, we definitely ain't in Kansas no more!

Chapter 13

"Everybody in life is a chameleon."

Melanie Chisholm

On October 1, 1975 what has been hailed as the "greatest boxing match of the 20th century" between Mohammad Ali and Joe Frazier was held at the Areneta Coliseum in Manila, Phillipines. Historically, this third and final bout has been referred to as the "Thrilla in Manilla." This was Ali's final answer to a long and open ended debate which was designed to squelch the controversy on deciding who authentically deserved to hold the title of Heavyweight Champion.

I was on board with Ali.

Everything he vocalized, I internalized.

Sold on his skills and his self-proclaimed declaration of greatness, I wanted to be just like him. I had always revered Ali and as a young woman, I felt that we had even more in common than we did when I was a little girl. Ali was fighting for his title and I was fighting for mine. He was trying to prove himself to a world that seemed to have forgotten that he truly was "the greatest" following a stripping of his championship title in 1966 when he refused to serve in the Vietnam War. As for me, I was doing exactly the same—dancing in my own self-created ring of greatness endeavoring not to remind the world that I was great but *prove* to them that I was. But it appeared that those around me wanted desperately to keep me boxed in their small versions of the world.

"I want you home when I get home," my peculiar roommate declared one day.

"What?"

"That's right," she went on to explain, "I don't want you roaming the streets . . . I want you here at home."

"Okay," I responded without resistance. Since she was paying the rent, I had little to offer by way of complaint, so I hunted for jobs by day and arrived back home in the evenings before she got home. However, in reality, I don't think Carmen wanted me to get a job. She wanted me penniless so that she could be the keeper of my destiny. But no one had earned that right—not my parents, not my brothers, friends and least of all—Carmen. Hence, I began to dream a dream again and ponder the possibility of life elsewhere.

Not better.

Not worst.

Just different.

One day while flipping through the Sunday paper, a small flyer fell out of the Classified section and onto my feet. In black bold letters the caption on the advertisement spilled out and poured itself all around me.

Free food.

Free clothes.

Free education.

Thirty days paid vacation!

"Damn!" I said out loud, "This is a deal I can't refuse!"

I immediately sought out the employer who was offering such exquisite benefits, and was surprised to discover that this gig was being offered by none other than Uncle Sam. He was recruiting for the military. However, up until this point, I hadn't thought about life in the quantities as the military as a viable option for my livelihood. The promises were carried in on piggy back hanging off the edge of large doses combining fairy tale and farfetched ideas that seemed perfectly reasonable to me—and all desperate people for that matter. The military was my ticket in! It was my way to ensure through the GI bill that I would get a fair shot at a college education.

My first stop was the Marines because they had the reputation of being the best. And the *best* was what I was interested in. However, upon my arrival to the Marines recruitment office, I was kindly informed that "the best" place for a woman was the Air Force.

"Okay," I said shrugging my shoulders. "Air Force here I come!"

When I arrived to the Air Force Recruitment Office they were singing the tune of my acceptance, however, I would have little time to learn all of the necessary notes required for participation. And in the summer of 1976, I received an urgent call from the Air Force with a message so strong that it changed the course of my immediate destiny.

"Cheryl," the recruiter began, "You better hurry up and get down here!"

"What?" I responded with panic in my voice.

"Yeah . . . they're about to cut out the GI Bill."

"Cut out the GI bill!" I hollered.

No GI Bill.

No college dream.

Welcome to the world of ordinary living for the rest of my life.

No thank you.

"I'll be there first thing in the morning," I promised.

Next thing I knew, Cheryl Waiters had signed on for a four year term with the United States Air Force and I was being shipped off to Lackland Air Force Base in San Antonio, Texas. I was scheduled to undergo eight weeks of basic training whereupon I would be transformed into America's greatest air man—though women were not granted permission to fly, which I always found ironic because the Air Force boasted of the sole mission "to fly and fight."

My initiation into this elite, sex-crazed and male dominated branch of the military began with the single act of being hauled into a room which held approximately one hundred new recruits—all men. They were young, testosterone-filled, abrasive and most definitely over-sexed.

"Get a good look at this woman!" shouted the commander. "It's the last one you'll see for a while!"

The men cheered and made obscene gestures, both with their eyes, tongues and ultimately with their minds. I could feel myself shrinking inside of my own skin as I was pranced around the room like some kind of show animal. *You have got to be kidding,* I thought to myself. *This is what I signed up for?*

Gender equality and consideration for women as equal recruits was not up for grabs in the Air Force back in the mid-70's. There was a clear distinction between male and female soldiers. The Air Force's mission of "flying and flight" was not extended to those without a penis. Women did not fly nor were they given an ounce of merit toward that end.

There were other clear distinctions of separation as well. For instance, the women didn't march as far as the men nor did they exercise as rigorously as their male counterparts. In addition, women did not shoot guns, go through the mine-field nor did they explore the gas chamber. Obviously, this touched upon the sensitivity of my childhood issues with regard to the preferential treatment of the male species and in simple terms, I found this aspect of the military to be repulsive. Between the gender bias and the sexual addictions of the sex deprived officers, I thought that the early phase of military life would most definitely drive me crazy. When women were not being relegated to what felt like inferior positions, they were constantly being propositioned for sex. *Everybody* wanted sex. The male soldiers reminded me of little more than dogs in heat searching for a warm nest to settle their manhood into. It was so pervasive that I wanted to design my own custom t-shirts with a simple slogan stretched boldly across the front which read: *PLEASE DON'T ASK ME FOR SEX.* The soldiers I encountered in the military world affected me deeply, but not all of its affects were negative based upon primal urges left unchecked and gone mad. My staff sergeant, a brilliant woman named Rosario, was one of the people who had quite a positive influence upon my life. She made a difference in my days of basic training. In fact, she made good in my life. Rosario had this amazing ability to almost peer into my soul and it had been a long time since *(if ever)* that someone had bothered to look so deeply into me.

"I know more about you than you know about yourself," she insisted. Her words stopped me cold and I believed her. "You can do life two ways while you're here," she began. "You can have self- discipline or imposed-discipline."

Interesting, I thought to myself.

"One is easy," she went on to say, "And the other much more difficult . . . but both are always your choice."

Needless to say, I opted out under the clause of "keeping my sanity" and chose self-discipline. This felt much less like torment to me and somehow on a wing and a prayer, the eight weeks came and went. During this time I would be humbled again and again, and I would also meet miracles and great surprises along the way.

One day while marching with my squad, we happened upon a squad of male soldiers and one man stood out to me. I don't know why he caught my eye, but nonetheless, he did and when we exchanged a brief glance. I nearly fell over when I realized that he was Ziggy!

Ziggy from Cleveland Ohio!

I knew that he had joined the service but never in a million years did I imagine to encounter him during my tenure in basic training. Perhaps we truly were soul mates after all. We didn't speak at the time for it would not have been appropriate to do so. But a single glance told me all that I needed to know.

Ziggy was changed somehow.

Slightly different.

Shortly thereafter, he got out of the service. Ziggy had problems. I didn't know what they were—but I just knew they were there. Perhaps he left to find answers to his problems. I guess we all have to find our own answers and our own miracles—which never seemed in short supply during basic training. I actually met someone whose mother lived across the street from a woman that I knew very well in Dallas. She wasn't just any woman—she was my *mother!*

Imagine that!

During the course of conversation, the woman asked if I was related to a Mary Strickland in Dallas, Texas. The question itself was shocking to me?

"Why do you ask?" I was curious to know.

"You look like Mary Strickland . . ." the woman said with a laugh.

"Yes! That's my mother!" I said with my jaw on the floor.

I was actually delighted to find out that she was doing well and living in Dallas. We spoke briefly on the phone in that moment with a promise of "more to come."

More to come, Mary.

More to come.

Upon graduation from basic training, I was initially assigned to the task of aircraft mechanic. I was offered the "crappy assignment" based upon my last name, which began with the letter W—so I was always at the end of the line since most rotations began with the letter A and worked backwards from there. The letter "W" was pretty much the last of the heap and the shittiest of everything.

"Waiters," I was instructed by my superior officers, "Pick up that wrench and handle it."

"What?" my shocked expression and bucked eyes seemed to suggest. The damn wrench was three feet long and weighed Lord only knows how much.

"Pick up that wrench and handle it!" they repeated, as if I didn't hear them the first time.

In that moment, I did what any self-respecting airmen would do—I burst into uncontrollable sobs, which in the moment was a sincere expression of my disappointment. But in the end, that expression may have saved me, at least temporarily. The Air Force altered my contract and instead of aircraft mechanic, I became a medical lab technician. I'm not saying that this was my dream job but it was a good stone throw's distance away from carrying a three foot wrench on my backside.

Following Basic Training I was transferred to Scott Air Force Base in St. Louis, Missouri. The year was 1977 and I was a wide-eyed, crisp nineteen year old woman who was ready to take the world on according to my own terms. But to my utter shock and surprise, I realized that the terms had changed when I landed on the steps of the airport with two heavy bags and a large black trunk. I approached a man seeking help with my luggage and was almost snapped in half by his bitter statement, "Hey! This is women's lib. Do it yourself!"

What?

This comment took me back and I had to take a moment and seriously ponder whether or not I was truly interested in living the long term effects of women's lib, especially if I had to carry my own damn bags. I knew in that instant that the world had changed *again*.

Upon my arrival to Scott Air Force Base, there was a period of adjustment in doing military life for a living. However, there were some things that I just couldn't seem to wrap myself around in adjusting to—namely having a roommate. Plain and simple, I didn't want one. But my preference in solo living conditions would be largely ignored, so I took it upon myself to create conditions that dictated that I would be alone at the end of the day.

In my observation of human behavior I discovered that white folks didn't like heat and black folks didn't like the cold. So, whenever I was assigned a white roommate, I turned up the heat to the point of intolerable under the fake assumption that I had a condition in which I was always cold. And when I had a black roommate, I did the opposite. I opened the windows in below freezing weather under the fake assumption that I was stricken with some kind of pre-menopausal hot flashes. Either way, it always worked like a charm and I consistently managed to stay untangled and free of roommates. Everyone who lived with me thought I was some sort of crazy and decided it was much easier not to be bothered with my internal weather changes.

But not everyone thought I was crazy and I did manage to make a few friends along the way. I befriended this interesting couple who always invited me to their place toting peace offerings of fine wine while they flaunted their lesbian love affair in my face.

"We're lovers," they used to boast.

"Oh," I said with one brow raised. This kind of situation would certainly draw the interest of superior officers because homosexuality was a big "no no" in the military. But this couple loved it. They seemed to gain enjoyment in living on the edge of sexual taboos, and I allowed them the pleasure of their intoxication and never breathed a word to anyone. They were kind to me and I was kind to them and at the end of the day that was all that mattered.

I learned a lot about life while living at Scott Air Force Base and had two close brushes with marriage. Given my cultural conditioning and predisposition to an aversion to the lifestyle, I was surprisingly open to considering a marriage proposal from Boy #1, who was a single dad who had sole custody of his one-year-old son because the mother had some "serious issues."

I like him well enough and we got along just fine, but I was profoundly turned off by his desperation to acquire a wife quickly. I got the feeling that he was in the market for an

"insta-mom" for his son and I just couldn't do the ready-made-family-gig. It didn't feel right to me so I bailed on the guy and his proposal.

Boy #2 was a Muslim. I had dealt with Muslims before so I was familiar with the breed and culture. I thought all would have been well till he took me to the Mosque and paraded me before one of its leaders who had the nerve to examine me as though I were a piece of cattle before granting Boy #2 permission to take me on as one of his wives! In fact, I would later learn that he had intentions of having more than one wife, but I was designated the head honcho of all wives.

No thank you!

Needless to say that I bailed on that one too!

I learned a lot not only about boys but also about myself while in St. Louis. One night while I was out on the road, my car broke down. Stranded and in the dark, I feared for my safety being a woman so I devised a brilliant plan to conceal my gender by hiding beneath a big, oversized coat and pulling a skull cap down low over my brow. I made it all the way home and no one bothered me. I felt a surge of empowerment by the act of being a chameleon and from that moment on, I knew that I could change myself—not from being a woman to a man, but from being weak and helpless to being powerful and transcendent. It was a valuable and much needed lesson that I would take with me into the very next chapter of my life—that of the ever changing chameleon.

Chapter 14

"Nobody can go back and start a new beginning . . . but anyone can start today and make a new ending."

Maria Robinson

In 1978 I was transferred to Travis Air Force Base in San Francisco, California. A city with its own unique heritage and lifestyle, San Francisco took its official name from the mayor in 1848 when the town was still a humble affair boasting of 469 residents. However, once gold was discovered at a local mill, it seemed as though the whole world poured into Frisco—swelling the tiny town to 35,000 inhabitants. The Gold Rush forever changed the face of San Francisco and put it on the map of international appeal.

Once I got off the plane and touched upon the ground, I could feel excitement and energy coursing through the land. This was a city where people came to dream about striking it rich and making good on the promise of wealth. Most noted for its diverse cultures, the Golden Gate Bridge, Victorian architecture, cable cars and the fog—it did not take me long to fall in love with this city by the sea.

As a Medical Lab Technician in the Air Force, to say that I felt under-stimulated was an absolute understatement. If the military was my key to the promise of a better future, I seemed to have taken a detour somewhere along the way. I felt far from the notion of being upwardly mobile during my tenure in the service. In fact, I oftentimes felt as though I were going backwards and then spinning around in a circle while trying to catch my own tail; but I kept an open mind and was always on the lookout for the unusual—which sat just beyond the borders of ordinary living. Cynthia, commonly referred to as "Cye" was one of those extraordinary things.

One Sunday morning day while preparing for a trip to Sacramento, I noticed a woman just outside of my window who was walking—no gliding, down the walkway.

"Who is that?" I asked myself out loud. I had never seen a woman who had such an interesting looking figure and an equally charismatic attitude. She had this beautiful black skin which seemed to shimmer under the sunlight and this incredibly long neck with the longest legs that I ever did see!

"Oh my God!" I screamed. "She looks like a horse!"

Those words may sound like an insult—but they weren't. She was majestic and I called her a "Black Stallion" because that is what she purely was. Beautiful in every way, she was a mammal with flesh for skin and elongated limbs—which moved with grace, elegance and the consistent flow of ease.

"I gotta know who this is!" I exclaimed as I raced from my bedroom and onto the walkway.

"Excuse me!" I shouted running after her and upon hearing my call, she stopped just like a prize winning Stallion upon command as she turned to me and smiled.

"What's your name?"

"Cynthia," she replied extending a hand, "but they call me Cye for short."

"I'm Cheryl."

"Nice to meet you, Cheryl."

Following a brief introduction on the tales of our lives, Cye and I quickly became friends. Six years older than me, Cye was filled with worldly wisdom and insight into the deeper meaning of existence. I was fascinated by her quick wit and charm, along with the knowledge she had acquired on the mysteries of life.

"Do you meditate?" she asked me one day.

"No," I replied, "but I probably should."

"I learned about Transcendental Meditation . . ." she offered.

"What's that?"

"It's a mantra meditation that takes the mind to a quieter place," she said, "It's the ultimate form of relaxation."

"Teach me," I quickly responded with enthusiasm.

The next thing I knew she was teaching me how to induce a transcendental state of mind through breath and sound. As a quick study and someone who always enjoyed running ahead of the class and mastering the lesson, I was no less interested in doing the same with this meditation. Though at the time I didn't quite know what I was engaging on a deeper level till one morning I was laying on the bottom bunk of my bed meditating. On this particular day, I was really getting into the zone—a place of mental stillness and detachment to my surroundings. The next thing I knew, I lifted right up out of my body and floated just above the bed before I realized what had even happened!

Oh my goodness!

I had induced an out of body experience which freaked me out! Terrified, I eased myself back down and like a soft blanket—poured myself back into my body. Needless to say, that was the end of my days of meditation. I was just too freaked out to try it again, but there were other experimental changes I was willing to make. One such change was the feminization of Cheryl Waiters. Upon meeting Cye, I was wearing an afro and was more of a tomboy—but that was all about to change.

"Girl . . ." Cye said, "We gonna have to gussy you up a bit."

"Not interested," I offered.

"What do you mean not interested?" she pressed.

"Beauty's commonplace . . ."

"Huh?" she asked with a raised brow. "I don't understand what you're trying to say, Cheryl . . . you're a very attractive woman."

"Yeah . . . but I'm not trying to be anything ordinary," I boasted. "I'm not trying to flaunt my beauty like the rest of these air heads. I'm taking the road less traveled."

"Come again," Cye replied in a sassy tone.

"I'm using my brain," I responded with attitude. "Beauty will get you dinner and some flowers . . . but smarts will get you a mansion and a yacht."

Cye burst into laughter.

She understood.

"Well," she said, "You can have your mansion and your yacht . . . and look damn good in both of them."

I laughed.

I, too, understood.

"Alright," I said surrendering my position. "Just a little . . ."

"Just a little," she promised.

Ultimately, we did a lot. Cye gave me my first perm and dubbed me the nickname "Cher," which I was proud to own. She also pierced my ears and was instrumental in transforming my wardrobe. After Cye's overhaul, I liked what I saw and made friends with the ambiguous gesture known as beauty. In the end, it turned out to be a wonderful friendship—the one I developed with beauty and the one I had developed with Cye. I looked up to Cye and accepted her wisdom on good counsel that she had lived longer than me and seen more than I had. It was Cye who told me that "don't nobody in this world really care about you or how you feel."

"What?" I questioned her as though I were a naïve child. Perhaps in many ways, I was.

"Nobody," she reiterated.

And I believed her—so much so that I cried profusely upon hearing those words and even more so after reflecting upon my life's experiences with my mother, father and brothers. *Maybe she was right,* I thought. *Maybe nobody really cared at all.*

Mama didn't care.

Daddy didn't care.

And the military damn sure didn't care. Up to this point, I had experienced military life to be sterile and void of the simple traits that made us human. And much like my relationship with my father, there seemed to be something essential missing in the exchange. When I was on duty everybody looked the same, walked the same and talked the same. Being in the service was like moving through the aftermath of an experiment in a sea of identical twins where something had gone terribly wrong and the womb of creation spit everybody out with an identical blueprint.

Yikes!

And as for women—they were definitely on the low end of citizenship in the military. In fact, they were more of a kin to cargo and each month saw the arrival of a new shipment. Once enlisted—you became government property and the military dictated every square inch of your movement. If they didn't authorize it, you weren't going to experience it. But there were certain experiences, in which military personnel was expected to become human guinea pigs. One such case would be the development of the "swine flu vaccination" which was tested on us first before its availability to the general public. I refused the shot and was nearly court marshaled for disobedience. But I refused to surrender my position as I had no interest in experimenting with *anybody's* vaccination, whether it proved beneficial or not.

There were tokens of the government's affection in every comment the superiors made. For example, "*if we wanted you to have a husband or a wife we would have issued you one.*" Sayings such as these went far in the translation of what it truly meant to be a part of any branch of the service. However, that's not to suggest that the entire experience was a wash. I acquired a lot of positive things from my time spent in the Air Force. I learned the value of discipline and mastered the art of self-regulation. As a member of the MAC (Military Airlift Command), I was afforded the privilege of travel and I enjoyed that immensely.

Ultimately, being in the Air Force boosted my overall confidence and offered just enough challenges in order for me to deliver the very best of me into the world at large. And for that, I will always be grateful. But this was sadly overshadowed by the fact that they simultaneously broke every promise ever made to me prior to my enlistment. At the very end of my training the Air Force decided to cross train me so that I could venture off into another unfulfilling career path—this time in Warehousing.

"Warehousing?"

What in the Sam Hell?

At this point, I was outdone and over it so I did something bold and outrageous—requested an early dismissal from the military due to my discontentment with their ability to uphold all that they promised to do. And believe it or not—they must have agreed because they turned me lose and set me free.

I was granted an early release from the Air Force.

Prior to my release, I had reconnected with my brother, Chuckie. He was enlisted in the army and living in Kansas boasting of a "fabulous life filled with a gang of opportunities."

"Cheryl . . . you need to come to Kansas," he persisted. "You'd do great here! There's so much opportunity!"

"I'll think about it," I always assured him. And after much deliberation and a California earthquake the decision was made in a flash. Before the ground could stop shaking I was on the phone with Chuckie, "Brother . . . I'll be there tomorrow."

And just like that I was on my way to Kansas in search of the Wizard of Oz, a brain, a heart and some courage.

It was a move that I still regret to this day.

Chapter 15

"Life is like riding a bicycle. To keep your balance you must keep moving."

Albert Einstein

Upon arriving to Kansas City, I learned some interesting things about the city. For starters, once upon a time it was against the law to serve ice cream on top of cherry pie and I couldn't help but wonder how they punished such an unforgivable violation. I also learned that the first female mayor, Susan Madora Salter, in the United States hailed from the state of Kansas back in 1887. This suggested progressive thinking, and it almost made up for the ice cream and cherry pie sin, but not entirely. But at the end of the day, that was neither here or there—what I was most concerned with was cashing in on all the promises that Chuckie had made on the phone when he boasted so vigorously about "the good life" that Kansas City had to offer.

"Hey Chuckie!" I said offering an embrace upon meeting him at the airport.

"Hey Cheryl!"

When we hugged, I felt an immediate disconnect.

Uh oh, I thought. *Something doesn't feel right.*

He felt more like a stranger than family, more of an acquaintance than a friend. More of a distant relative four times removed than my first sibling and brother. This feeling only expanded when I returned to Chuckie's place and saw with my own eyes that his self-proclaimed declaration of living "on the high horse" had been mildly exaggerated to say the least.

"It's good living here in Kansas," he continued on his rampage, but for the life of me I couldn't see what was so good about it. In fact, from all that I could see—or rather *couldn't* see surrounding his claims of a "fabulous life," I chalked up his fanciful tales to two things: 1) fantasy and other . . . falsehood. Disappointed and depressed, I knew that I had to do something with the rest of my life, so I got busy in the pursuit of higher education and with the help of the GI Bill, I enrolled at Kansas State University.

I began with a curriculum of general studies and within time, I eased into the groove of the city and its flow. "I can still make something of myself," I was convinced. And that consistent thought seemed to ease my mind.

I can still do it!

I can still be SOMEBODY!

I can make a valid contribution to life!

I can do anything!

And this was my mantra till two intimidating drug dealers stepped onto campus one day and pinned me in a corner, slightly changing my sunny disposition to a droopy desperation.

"We're looking for Chuck . . ." they barked in no uncertain terms.

"Chuck who?" I withered while asking. "My brother Chuck?"

One of the goons nodded while the other cracked his knuckles.

"He ain't here . . ." I responded humbly. "He doesn't go to school."

"Let him know we got unfinished business . . ." said goon one.

". . . that needs to be *finished*," the other goon chimed in, completing the sentence with an exclamation point of his own.

I nodded.

When they left, I bolted home and called my people back in Cleveland.

"Chuckie's in trouble! Chuckie's in trouble! We gotta get outta here!"

"Wait! Slow down! Uncle Willie demanded on the other end of the line. So, I paused for a breath before quickly jumping back into the story about how these thugs came looking for Chuckie at the school.

"They don't mean him no good," I told Uncle Willie. "And something real bad is gonna happen if we don't get out of here." But something bad had already happened—unknown to me at the time, Chuckie had been dabbling in drugs and all things unholy. He had been keeping the company of unsavory men and was pushing and using drugs on the side.

This was his so-called fabulous life?

I felt duped.

Again.

Ultimately, Chuckie's story read like a bad episode of a soap opera that didn't make it past the first season before being cut and thrown off the air forever. He had taken a fancy to dating a captain's daughter—which was not well received by the captain, who instructed Chuckie to "stop seeing her" at once. But Chuckie, being who he was—didn't take too kindly to such a suggestion so he kept right on about the business of bedding the captain's daughter—till one day hell unleashed its own fury and a fight broke out between the captain and Chuckie.

Guess who lost?

You guessed right—Chuckie. He was court marshaled and shipped off to Ft. Leavenworth prison where he was dishonorably discharged from the service.

Well, so much for the good life.

It was time for me to take another detour.

Bye bye Kansas.

I found myself back in Buffalo and standing on the doorstep of my former stepfather, Thomas Strickland. After a brief phone conversation with him, I was worried about his well-being. He didn't sound right on the phone. When I re-entered his life, I saw that he had been frozen in time and was a mere shadow of the progressive and upwardly mobile man that he once was. Unshaven and unkempt, he had been reduced to almost nothing in stature, size and spirit. The house was in shambles, the roof was leaking and the ceiling was caving in.

"Tommy . . . what's going on?" I pleaded.

"She's coming back . . ." he whispered. "She's coming back."

I couldn't believe it.

After all of this time, he was still waiting for my mother to return. He was staring at the front door as though he were expecting a ghost to come prancing through the doorway.

It was eerie.

"Tommy . . . she ain't coming back," I said. "She's moved on with her life."

He did not seem capable or willing to take in those words. He was stuck on the ramblings that a mad man tells himself in a desperate attempt to believe the unbelievable.

"What?" he replied, stumbling over my words in disbelief.

"She ain't coming . . ." I repeated.

He slumped in his chair without responding.

"Tommy?" I called out to him.

Silence.

"Tommy?" I repeated.

And still silence.

He was so broken that it pained me greatly to look upon him in this condition. He had done so much for me and my family when I was a little girl and somehow, I wanted to repay him. Therefore, I decided to stay a while and help Tommy breathe new life into his tired body and withered spirit. Change was in order and it desperately needed to be made. One evening after Tommy had left for work on the graveyard shift, I felt the presence of subtle energy in his house. It did not feel welcoming or warm. I got the feeling that something *other* than Tommy and his pitiful pining over my mother was happening in this home. That night, I wrestled with the unknown. I called forth love to replace the painful energies that I felt surrounding this space and taking up all of the breathable air. It was a long, dark and restless night and by the time dawn broke, I was exhausted—but somehow *free* of what had lingered there prior to my arrival.

Interestingly, that morning when Tommy got off of work and he opened the door, the sun was shining brightly against his back. His face was radiant and his familiar smile had returned. He was in a great mood.

"Guess what?" he said enthusiastically, "I got a new apartment!"

Tommy was free. It was the beginning of something new and wonderful. Ultimately, he made the choice to go on *without* my mother. It looks like Einstein was right after all—to keep your balance you must keep moving. Balance had been restored to Tommy's life and his affairs. And in keeping with the spirit of progression, I, too, was ready to move on and try my luck with the navy.

The year was 1980 and I was twenty two years old.

"Cheryl . . . you should go and visit with your mother," Tommy insisted.

"But I was thinking about joining the navy . . ."

"Go to Dallas first," he said.

"Okay," I agreed with a heavy sigh. Every time I went to a new city, I always ended up with a new destiny.

Wanted or unwanted.

Welcomed or unwelcomed.

Desired or undesired.

But new nonetheless.

And Dallas was no different.
Wanted.
And unwanted.
Desired.
And undesired.
But new nonetheless.
Again.

Chapter 16

"When we are no longer able to change a situation . . . we are challenged to change ourselves."

Victor Frankl

Next stop . . .

Dallas, Texas.

It was a new chapter in my life, though it seemed to have an odd familiarity to the old chapters which had already been written. Alaska was the largest state in the United States and Texas ranked number two on the list boasting of twenty four million people and sixteen million cattle. *That's a whole lot of cows.* I guess it goes without saying that a vegetarian diet is probably *not* high on a Texan's list of priorities.

The state of Texas supplies more wool than any other place in the nation and it is also noted for the largest herd of whitetail deer. The famous soft drink, Dr. Pepper, was invented in Waco, Texas in 1885 and Dallas is home of the frozen margarita machine. Pity, I didn't drink . . . I probably should have taken it up as a hobby by now.

My mother was living in a two bedroom apartment in the heart of Dallas. When I arrived at her doorstep, a pit settled in my stomach as I waited for the door to be opened.

Knock!

Knock!

Who's there?

Nobody answered.

Knock!

Knock!

Who's there?

I had begun to feel as though I were in the middle of a bad "knock knock" joke.

Finally my mother appeared.

"Cheryl! Come in!" she squealed upon opening the door. It had been some time since we laid eyes upon one another. She embraced me with intensity—as if she were trying desperately to bridge the gap between the distance of she and I. As I stood on her welcome mat, I could not help but wonder if we held on to one another long enough, could we melt away a lifetime of misunderstandings, intentions *and* inflictions? Was it too much for the little girl to ask that the grown woman be healed too? It was a new day and I wanted so desperately to believe in my mother again. I wanted to extend grace in the places where none had felt possible up until now. And even still, it might have been just too much to ask of the moment.

"Come on in," she repeated, as if I might turn in the other direction and run off with a quick change of my mind.

Upon entering her home, my heart sunk.

The place was a dive.

A zero.

"Thanks," I responded as I looked around at all corners of the room which all bore the same theme—*poverty*.

"What's really going on here?" I couldn't help but ask. I had never seen Mary living in such substandard conditions. Her lifestyle had diminished greatly and she seemed to live just a hair above poverty, if at all. Michael and Keith came running out of their bedroom and I was shocked when I saw how big they were.

"Oh my God!" I said, "Look how you guys have grown!"

"Cheryl!" they said running into my arms. Like little soldiers, I couldn't believe how tall these little men were. In my younger brothers, I saw just how much time had passed and how greatly things had changed in between the beats of time. While Mary was still a beautiful woman, she looked tired and worn—as if a lifetime of struggle had caught up to her and made its official stamp on the essence of her being.

"How you doing, Mama?" I asked, while I listened to all that she did not say between the pauses of "fine, good and swell."

"You sure?"

"Yeah I'm sure . . . why wouldn't I be sure?" she put the question back to me. I could tell the inquisition made her uncomfortable—because she knew just as I knew. She looked to be a long way from fine, however, I didn't linger in her misery because I had a host of plans for the day, which included spending time with Mary's father—my grandfather, who was a red-skinned Indian man.

When he came and picked me up, I was surprised by the beautiful red tone of his skin. In some ways, he seemed to be something other than what we were. He was kind, considerate and very open to playing tour guide for the day. Graciously, he scooped me up in his automobile and took me on a tour of the "big city," which resembled something more like the back woods.

I couldn't help but notice that the locals shuffled when they walked. No one seemed to be in a hurry to get anywhere at all—in their day *or* in their life. I found this to be common just about everywhere we went. From the grocery store to the corner market, everybody seemed to be just "bobbing" along—weaving in and out of the tapestry of their lives without complications.

Welcome to the South.

"How ya'll doin'?" was the standard greeting. And everywhere that we went people were more than willing to engage my grandfather in conversation—albeit the slowest, moving conversation you ever did hear.

"How you doing . . . ?"

"I'm getting' along mighty fine."

"How's the wife?"

"Wife is fine."

"Who's this?"

"Granddaughter Cheryl . . ."

"Hi Granddaughter Cheryl . . . how you doin'?"

Okay.

Just for the record.

I'm definitely a city girl.

After spending the day with my grandfather, I was delighted to see that he was so well respected in the community. He had his own way about him which I loved and embraced, and from the looks of things, so did everybody else. Maybe Dallas wasn't such a bad pit stop after all.

I got to thinking that I could do some good here and help Mary out.

"What do you think about sending the boys to Tommy," I proposed, "and me moving here for a while to help you get on your feet?"

Surprised by my offer, she paused a moment in contemplation.

"Okay," she said without resistance.

"Okay?"

"Yeah . . . I'll do it."

And just like that Mary and I were back together again. She shipped my two younger brothers to Buffalo to live with Thomas and we moved to North Dallas. I wanted to help elevate Mary's lifestyle and relieve her of the burden of single parenting so that she could get herself together.

"Why don't you go back to school?" I asked her, only to find a pile of excuses in response.

Too old.

Too tired.

Too poor.

Too busy.

But there was one thing that Mary was never too old, tired, poor or busy to do—hit the club. But sometimes you have to know when it's time to "let go" and move on about the business of your own life. And I couldn't stay "stuck on pause" trying to help elevate someone who had already reached their greatest potential—at least in their own eyes. Therefore, on behalf of my own dreams I enrolled in El Centro Community College. At the time, Texas had the poorest education system in the country and this would present its own set of challenges in my slow climb to the top of the food chain. In pursuing my engineering degree I was required to take Calculus—which I was shocked to discover was a "self taught" class.

Self taught?

This wasn't Basket Weaving 101 . . . it was Calculus!

What in the hell?

In the meantime, I found another experience to experience which changed my life forever—once again. My life was funny like that. It changed on a minute's notice, but most of the time I didn't even get the whole minute to prepare for the change.

One night while visiting the local library I came upon a magazine which caught my attention. The caption on the front of the magazine read, "America the Beautiful . . . America the Condemned."

"What's this?" I asked myself drawn to the cover. After a bit of investigation, I discovered that the magazine was published by the Worldwide Church of God. I sat down, began reading and became so engrossed in the material that the next thing I knew, the lights were being turned out in the library.

"We're closing," said the librarian, jarring me from my self-induced trance.

"Uh . . . okay," I stammered as I quickly gathered my things. When the librarian walked away, instead of placing the magazine back on the table, I shoved it in my backpack. I quickly made my escape with the "material" in hand as I ran through the night streets like a high profile cat burglar. Later, I looked back on this and I laughed . . . because when all was said and done, the religious magazine was *free.* And ultimately, it doesn't get any funnier than stealing something that's "free."

After this encounter, I became fascinated and obsessed in learning more on what the Worldwide Church of God had to say on just about every topic. As with all things in my life—I dived in head first.

I took the Nestea Plunge into the church and converted to become one of its most zealous members.

Not only did they also keep the Sabbath, but they observed all of the Holy Days. Studying at the Worldwide Church of God was so profoundly educating it was like being enrolled in a Bible University. In total, I read the Bible cover to cover fourteen times!

According to the Worldwide Church of God, I was shocked to discover that the crucifixion happened on a Wednesday and not on a Friday. Upon learning this, I went straight to see the minister at the Seventh Day Adventist church *(where my current faith was placed)* and inquired about my new findings. In short, the inquisition was a bust and I left the church that day never to return.

As a diligent student of the teachings of the Worldwide Church of God, I took my new truths to heart.

I always did.

And like a sponge, I soaked it up and felt the quenching of a thirst that I had for lifetimes. The Church emphasized family—and under the weight of these teachings, I began to ponder my own family, especially the relationship that I had (or didn't have) with my father. I began to consider my roots and heritage, and although I had lived all over the country, Cleveland was where I began. I had spent the last two years of my life in Dallas and it was time to return to the land of my people—the "Cleaved-Land" and meet up with Charles Waiters for a face-to-face and get all that was "owed" to me.

I was intent on collecting *all* that was long overdue and coming.

Nothing could stop me.

Nothing.

And so without further ado, I bid farewell to the party girl Mary, closed the chapter on Dallas and headed back home. To this day, it was the worst mistake I *ever* made.

Chapter 17

"The idea of my life as a fairy tale is itself a fairy tale."

Grace Kelly

At the age of twenty four, I was headed straight to Cleveland with a message and a motive. I was en route to collect on all that was long overdue me—love, emotional support and a long list of apologies. "From whom?" you may ask.

"Just about *everybody,*" I would answer.

Upon arriving into town, I landed on the doorstep of my brother, Chuckie's house.

"Hey Cheryl," he said to me offering a casual greeting upon opening the door.

"Hey Chuckie," I replied.

Upon entering the house, I couldn't help but size up his entire living situation in a single glance.

"So . . . you're back," he laughed with a giggle.

"Yeah . . . I'm back."

I shook my head as I watched him maneuver through the complications of living with his girlfriend, a woman who had three kids of her own. In that moment, I felt disappointed in my brother. He had the potential to be so much more than what he had aspired to. He seemed to live one notch above "existence," taking up residence with a woman that his interest in seemed minimal, at best. It was almost as though he were just a body that was taking up space without sincerity, passion or compassion.

"So what brings you back?" he asked.

"I got business here," I stated flatly.

He laughed again. His humor seemed more demeaning than consoling.

"What's so funny?" I asked.

"Only thing that brings somebody back to Cleveland is a stroke of bad luck," he boasted. "So what's your story?"

"No story," I said. "I'm back here to do some things for myself."

"The best thing you can do for yourself is to get the hell out of town," he said with another laugh.

I ignored his poor use of humor as well as his advice. He seemed to be the last person on Earth who was in the position to dole out any sort of guidance to lost souls—since for all intense purposes he was at the top of the list. With that being said, it didn't take long for the Cleveland drama to kick in. I hadn't been in town long before I was bombarded with the bitter news that my brother, Chuckie, had been badmouthing me all over town. I felt betrayed and saddened by the lies and half-truths that he was spewing to the neighborhood and to the rest of the family.

Now, why do you wanna go and do that for?

His allegations and back-stabbing ways made me want to retreat and stay away from him altogether; but ultimately, his words didn't hold water because his own life was in such shambles. I doubt that many people took him seriously and within a month of my return, Chuckie had left his girlfriend and moved onto another woman. She was a beautiful girl named Cassandra—and what she saw in my brother in the form of a "viable" dating partner was a gigantic mystery to me. But when it was all said and done, Chuckie went on with his life and I went on with mine.

I had been introduced by Chuckie's girlfriend to a rather intriguing man. His name was Jerome Burkes. He was a different kind of man. I don't know if you'd call him a "good catch" or not but he translated well on paper. Jerome came from the kind of stock that people would call "good breeding." He was the offspring of a well-do-family who had an affinity for politics and upscale social events. In many ways, he was the man about town. And upon beginning a love affair with him *(if you could call it that),* I became the woman about town—attending a host of political events and social affairs.

This was grand exposure to the kind of life that I always felt well-suited for living.

I could be *That Girl.*

Financially, Jerome was good to me, but behind closed doors he was a dick. Excuse my French, but for lack of a better "anatomical" part to describe him, I can't think of a single adjective more suitable to define his behavior. In short, he drank too much and was a womanizer with an insatiable sexual appetite which could not be satisfied by just one woman—or so it would seem. Ultimately, this obsession preempted an understanding between me and Jerome and it looked a little something like this:

During our courtship we practiced the "rhythm method" as a form of birth control, which meant there were several days during the month in which we were abstinent. Under such restrictions, Jerome swore up and down that he could not go without satisfying his "animal urges" for an extended period of time; so I agreed to allow him to have another lover during the times in which we were abstinent. But with one stipulation—I was given the privilege of selecting the woman that he was approved to sleep with. Obviously, my primary interest was in selecting women who were clean and not "infection ridden." I had a keen eye and it was a good deal for Jerome. At the end of the day he got "two chicks" for the price of one.

He got his cake and ate it too.

It was every man's dream . . . or was it?

So imagine my surprise when out of the blue he announced one day, "Cheryl . . . I want to get married."

Say what?

He may have been good to me financially, but he damn sure wasn't good *for* me emotionally. Therefore, I was forced to re-evaluate our entire living situation when he "popped the so-called question." Living with Jerome gave me my first taste of married life and from what I could gather—it definitely wasn't the gig I had been searching for.

Think I'll pass on this one.

Beyond his selfish ways, womanizing and a drinking problem, I was privy to the *real* reason Jerome was so interested in marriage. He had run into some trouble with the law and was looking at doing jail time, so he needed to secure "our deal" to make sure somebody was left behind to look after his house and dog. This little situation was definitely not in my future's forecast, so I packed my bags and hit the road. My departure was a more elaborate plan than the words suggest and I would do it greater justice to call it by its real name—"The Great Escape" as it was just that. While Jerome and his "secondary girlfriend" made love in the bedroom next door to mine—I gathered up all of my belongings and got to stepping.

"I'm outta here!" I said to myself, but not a word was uttered to Jerome.

I can only imagine his shock and dismay when he discovered that at the end of his rendezvous, his "real" girlfriend was really gone. And interestingly enough, when I went down to the lake to clear my head from the drama of it all, I saw a familiar man walking towards me carrying a fishing pole. His eyes widened with great recognition upon seeing me, but there was a little room for doubt.

"Are you Cheryl?" he asked.

"No . . . no I'm not," I quickly replied and just kept walking.

I didn't pause, skip a beat or dare to turn around.

It was Ziggy.

I knew it was him and deep down inside, he also knew it was me—but he hovered just above doubt long enough to give me an out and I needed an "out." There was no way I could be reeled back in to Ziggy's madness on genetic breeding and the like.

After the breakup with Jerome, I started all over again.

Immediately.

I needed a quick "hide out" so I moved into Chuckie's building and rented a humble apartment. It wasn't fancy—it was simple just like the rest of my life. On the first night in my own place, I felt the weight of the city caving in on me. There was an internal pressure against my chest of a life that was missing the course of its full potential. I had so much promise and so much more to offer—I just needed the *opportunity* to deliver my goods to life. I fell to my knees and sobbed, asking God for help.

"Please help me . . ." I cried as I lay close to the ground. "Please."

Following that day, I began to lay the roots of my spiritual foundation.

Again.

I began faithfully attending the Worldwide Church of God. And after what seemed more along the lines of an FBI interrogation than a simple "meet and greet," I was granted membership and access into the church. After all of that, I threw myself into the Word for solace, comfort, guidance and direction. And that appeased my soul—for a time.

I cashed in on my G.I. Bill (which was meager) and utilized the funds to finance my tenure at a local Community College. Life began to stabilize and alas, I was back on track. But the hills, valleys, peaks and mountains are many—in life. And just when you think you're on track—

Bam!

Boom!

A detour.

The Worldwide Church of God's founder, Mr. Armstrong, died. And following his death, it seemed as though the whole church went to hell in a hand basket *(no pun intended)*. The original teachings became void and the rituals that made this church "unique" and special were abruptly set aside. We began to celebrate Christmas—which was not initially observed. We stopped keeping the Sabbath and began to hold services on Sundays. So much was changing so quickly—that I began to feel uprooted. My entire foundation was being "shook, rattled and rolled" according to whoever held popular court that week.

Where do you go when you believed that you had already arrived?

Start from zero and pick a new destination.

But where?

Being in church was no different than being on the street—prejudice began to take form and show itself on the faces and in the peculiar and odd mannerisms of the congregation itself. White members didn't want to shake black member's hands and vice versa. We became a congregation divided amongst itself on the basis of hate. This led me to question the entire basis of Christianity itself. Was this the message that Jesus taught or was his Brothers in Christ just plain 'ole crazy? Needless to say, following this adventure I began my own spiritual quest. Unfortunately, it was a solo journey.

In June of 1986 at the age of 28, I graduated from Cuyahoga Community College with an Associate's Degree in Mechanical Engineering Technology. I was the first in my family to graduate, and felt stupendous in that I was finally showing the signs of a greater promise to come.

In the fall of 1986, I transferred to Cleveland State University. While taking an engineering class, it was here that I was shocked to discover prejudice in its purest form. I had never seen so much ugly lived "out loud" on the faces and in the dirty, subtle deeds of mankind.

Were people really rotten on the inside and simply washed clean by bars of Ivory soap on the outside? It would seem that with all of the subtle and not-so-subtle racist remarks, statements and actions—that mankind would have blazed a trail of stench that stunk as far, deep and wide as the high heavens itself. Everything that I had ever been taught about people being "good and decent" was null and void. Cinderella had dropped dead of a massive heart attack and nobody had bothered to mention it to me—at all. In the simplest of terms, ladies and gentlemen . . . *the fairy tale was now over.*

Chapter 18

"We are the forgotten generation . . . I feel like an alien on earth."

Cheryl Waiters

Once upon a time ago, it seemed as though the world had turned their attention to one entertainer and the focal point of the question that everyone seemed to be asking was, "*what's wrong with Michael Jackson?*" To the world around him, it appeared as though right in the middle of an extraordinary life and legendary career—that he up and went stark raving mad. It was insanity at its best confined within the body of a man who possessed more raw talent in a single strand of his DNA than most would ever possess in a lifetime.

What was wrong with Michael Jackson?

Interestingly, the same thing that was wrong with me.

I was born in the same year as Michael and ultimately became a product of the same era. He and I grew up under similar generational experiences. We both breathed the same air and bought into the propaganda that America was selling to black babies born in the late 50's. We had been taught, shape, molded and brainwashed into believing in a world that no longer existed once we grew up. And there was no consolation prize offered to those who had played the game, believed in the hype and planned their lives around the great bubble of fiction that we had been tricked into believing. After the Civil Rights Movement had ended, racism was still alive and well. The only difference was we had to face a new form of cleverly "disguised racism" and at the end of the day the joke was on us.

Michael and I were one in the same and what was wrong with Michael was the same thing that was wrong with me. We were the forgotten generation and in many ways, confined to life on Earth as aliens disguised as human beings—when we felt like anything but human at all. Only to have the masses call us "freaks" and ask what's wrong with him *and* her?

In 1987 and at the age of twenty nine, I was birthed into the real world. It was a slow descent into madness or hell. Cleveland State University felt more like the University of Hard Knocks. Subtle and not-so-subtle forms of racism were emerging everywhere—and I began to see life through a new set of eyes. It was as though a veil of deceit had been lifted and what remained was the blatant truth about a world where I no longer belonged and, ultimately, never fit in the first place. The world was black and white—literally. Interestingly, even Michael Jackson would one day compose a song titled, "Black or White."

Welcome to the rest of my life.

It was yet another time of great transformation. This was a time of great blossoming whereupon I learned about the great Black men and women who helped to build this country and make it what it is today. While at Cleveland State, I pledged Delta Sigma Theta, a sorority founded in 1913 by twenty two collegiate women at Howard University. The sorority was built upon the principles of utilizing academic strength to be of service to people in need. To date, Delta Sigma Theba is the largest African American Greek Lettered organization in the world.

By 1988 and upon entering the thirtieth year of my life, I began to look upon my hometown with great despair. Cleveland, as I have often referred to as the "cleaved land" had grown into a segregated city that was as dirty on the outside as it was on the inside. The city was the brunt of everybody's jokes, and it did appear that the city had well earned its reputation. Hell, I couldn't defend the withering town's reputation, even if I wanted to. And in the midst of change, chaos and transformation—I hit yet another wall. And this time in the midst of an awakening and greater awareness of what was really going on, I was also hit with an identity crisis and felt as though I were spinning out of control on the edge of a single unresolved question: *where do I belong?* And where do I begin to address such a big question that would lead me from one end of the Earth to the other—ironically while standing completely still.

Time did not pause out of great consideration of me having a crisis.

Nope.

It kept moving.

In fact, it always does.

In the summer of '88, I accepted a position designing steel mill equipment with an engineering company in Cleveland. To my shock, the wages were meager at best.

"What happened to all this money that engineers are supposed to make?" I posed to a fellow co-worker during a casual conversation one day over lunch. In that moment, he laughed so loud that it almost ruptured my ear drum. I was stumped by what seemed to be an ironic display of outrageously uncontrolled laughter.

"What's so funny?" I questioned him.

"You won't make any money doing this gig . . ." he insisted.

"What?" I responded with bucked eyes. "What are you talking about?"

"Engineers don't make great money," he repeated himself.

I almost fell over at the thought of a life of poverty—especially since wealth was in my DNA, courtesy of Charles Waiters.

"Come over here to the window," my co-worker suggested.

"Huh?" I asked, looking down toward the ground which was several stories below us.

Hell, maybe he was gonna push me over the edge and save me from the misery of it all.

"Just come over . . ." he prompted.

I walked slowly to the edge of the window and looked down below.

"What?" I asked, noting that there was nothing to see on the ground except a large crane that was being used for the construction work being done down below.

"See that crane?" the co-worker asked me.

"Yeah . . ." I replied, foggy from his line of questioning.

"Construction is where it's at!" he blurted.

"Construction!"

"Hell yeah!" he insisted. "Construction workers get paid the big bucks!"

I put my hand over my head, which almost seemed to suggest the obvious—*I could have had a V-8* instead of doing penance and hard time devoting my entire life to a career that would leave me broke, busted and unhappy.

Ooooooohhhhh!

Now you tell me!!!!

Deflated, I returned to my desk to rehash every missed turn and mistake I had ever made along the way. My thoughts were like a runaway freight train carrying a full load of shit and no working brakes.

I should have stayed in California.

I should have stayed in the Air Force.

I should have gotten married and had children.

No, wait! We would have only ended up divorced and then I'd be on welfare!

Perhaps it was time to reflect on a new direction in my life—but where?

In the summer of '89, I was in the market in search of a new opportunity. I landed an interview with Keith Finch, the President of a progressive African American Mechanical Engineering firm in Cleveland. I was hope-filled and inspired by the opportunity to work for his company—but no sooner than I hit the door did I also hit the "floor" (at least my heart did) when Mr. Finch shot me down as a potential employee of his firm.

"Well Miss Waiters . . ." he began, "I'm impressed by your background but I don't have a need for your specialty at this firm."

"Okay," I wanted to say to him in response, *"No problem . . . if you could just help me pick up the rest of what's left of my shattered heart up off your floor, I'll gather my things and be on my way."*

"But that doesn't mean you can't have a bright future . . ." he said with a lot of hope in his voice.

"Really?" I asked, perking up.

"Plumbing is the way of the future . . . and there's a lot of money in it."

"Plumbing?" I asked as my voice tapered off dramatically.

"Plumbing," he confirmed.

"Well . . ." I sputtered, "I'm not a plumber, but I do know about electrical work. My uncle had his own business and he taught me everything I know about the business today."

"Hhhhmmmmmm . . ." he mumbled as he sat back in his chair in quiet contemplation.

"Hhhhmmmmmm????" I regurgitated for clarification.

"I got somebody I want you to meet . . ." he said with a big smile.

And just like that—I was sitting in front of Terri Hamilton. Terri was the director of a program called Home Town Prep, which was a company that had been established to help minorities build careers in the construction field.

"Cheryl!" Terry said with great familiarity in her voice upon meeting me. Since we had never met, I assumed her overly-friendly tone was indicative of the fact that she had heard good things about me *and* that she was definitely going to help me. And if she couldn't, she would find someone who would—which was where Barbara Fluellen came in handy. Barbara also worked closely with Terri and between the both of them, some hard core questions and a couple of intense "stare down" matches—Barbara was ready to make a proposal.

"Cheryl . . ." Barbara asked quietly, "What do you know about Local 38 Electricians Union?"

"Local 38?" I asked.

"Yes, the International *Brotherhood* of Electrical Workers?"

I paused, contemplated the question, and then stated the obvious.

"Not much," I said.

"Not much?" Barbara confirmed.

"Nope."

And I should have left it that way.

Chapter 19

"And if there was ever an hour whereupon I needed to be more than a girl, more than a woman, and definitely more than me . . . that hour had surely arrived."

Darnella Ford

The very next day I was on my way toting a one way ticket to hell. I was set up to meet with Local 38. First, I was given an agility test to determine how well I could use my hands. I passed with flyer colors, so I could move on to the next phase of the interview process. I was escorted into a conference room along with my representative, Barbara, who accompanied me to the interview. Upon entering the room, there were three men seated around the table—a contractor and two representatives from NECA. The NECA representatives were white men who wore stiff suits and looked like they belonged to the CIA. The air in the room was so thick you could cut it with a machete. Intimidation was the name of *this* game. Once the interview process began, these "CIA" agents started firing off a series of questions so fast that I could barely answer one before another one was coming in its place.

"Do you think you can be an electrician?" was the first question.

"I don't know," I fired back. "Do *you* think I can become an electrician?"

The room spills to silence—but only for a moment. They clearly recognize that I'm no pushover, nor am I so desperate that I would lose myself to acquire this position. Once noted, the interview continues with more questions and greater intensity.

What makes you qualified to perform such duties?

Why would a woman want to work on a construction site?

Can you handle the pressure?

What are your qualifications?

Why do you want to be an electrician?

I felt like a sitting duck in front of a firing squad. These were hard ass men and I was two steps away from sinking beneath the floor—just to get the hell out. This was intense beyond measure. I hadn't bargained on this kind of party. It felt as though the walls were closing in around me and I was being suffocated by a thick blanket of racism. *What kind of organization was this?* I couldn't help but wonder.

The FBI?

The Mafia?

After the interview ended, I walked out into the lobby with Barbara, I turned to her and asked, "What in the hell was that?"

In short, *that* was the beginning of the end. But when it was all said and done, I impressed the hell out of them. I stood to the full measure of whatever it was that I was supposed to be. They also fell in love with my military background and engineering degree. *I was born to do this*—and all of my advance preparation seemed to suggest that very thing. Needless to say, it was a slam dunk. After the rigorous interview process, I wanted to call Uncle Willie and ask him about Local 38 and why he wasn't in it. I also wanted to explore the heavy vibe of racism that I felt in the room—which seemed to dominate my interaction with the men.

What was really going on? Was this the infamous beast known as institutional racism, defined as "any kind of system of inequality based on race?" **Wikipedia Encyclopedia**

Was this what Black Power Activist, Stokely Carmichael, meant when he coined the phrase "institutional racism" back in the late 60's? I had wanted to put a call into Al Sharpton and become better informed on the nature of this beast. However, I would never get the chance because the Union swooped me up in three days.

Welcome to the party.

In August of 1989, I became an official member of Local 38's International *Brotherhood* of Electrical Workers. Out of two thousand members, there were approximately thirty women in the Union when I joined. And less than 10% of all members were minorities—so in essence, that was a double whammy for me.

Black.

Female.

And an outsider in a white male dominated fraternity.

It was raw.

I was done and out done both at the same time, but I just didn't know it yet. Interestingly, in 1978 president Jimmy Carter mandated a minimal goal which dictated that all Federally funded construction projects committed to hiring at least 6.9% women as part of the work force. Therefore, out of every one hundred men on a project, there had to be at least six female employees in equal capacity. Perhaps, this was where the real nightmare began.

I would later discover that Uncle Willie had never joined the Union because in his day, the Union bypassed minorities and shunned their participation and membership. But by the late 80's things were *(twenty five years following the Civil Rights Act)* supposed to be different—or so they said. However, it was little more than smoking mirrors and grand illusions. And that I would *soon* find out. But for the moment, I was on top of the world.

Legit.

In the beginning, joining the Union was the ultimate high for me. It almost seemed like the next logical step in the sequence of unfolding events in my life. I had always bucked the system by challenging limiting concepts of what a woman could do and be in our society.

In many ways, I came into this life with my "dukes up." As a child, I saw that being a boy garnished special favor, so I tore down the gender walls and did "boy things" to win love and approval. In the words of an old Peter Cetera song, "We did it all for the glory of love."

I broke all the rules on being a girl. I rocked the boat as the only girl in an all boy family. I shook things up as Air Force recruit. I pushed the limits as an engineer major and damn near tipped the whole thing over by joining the Brotherhood. *I did it all for the glory of love!* And joining the Union was like hitting the "mother lode" in a long list of accomplishments and achievements, which broke through the barriers and restrictions of being a woman.

Upon acceptance into the Union, I was required to fill out a stack of paperwork. Upon completing the forms, I was given a work order for my first assignment, along with a voucher and a list of tools that I would need to fill my toolbox. *Wow! This was actually more exciting than I thought it was going to be.* I felt like G.I. Jane and was ready to conquer the world. I made my way to the hardware store where I surrendered my list and the clerk filled the order for me. In no time, my empty toolbox was full. I had officially traded my briefcase for a toolbox.

Okay.

No problem.

The next day, I was sent to my first job. As instructed, I reported to an electrical trailer armed and ready for work. Prior to my arrival, I was instructed to wear jeans and boots—which I did. My boots were flat, black and rode up to the knee. In my estimation, these boots were sufficient enough to handle the "dirt and mud" of a construction site. But I would soon learn upon my arrival that I was sadly mistaken.

When I opened the door and walked inside of the trailer, I immediately spotted two, unfriendly looking guys—a black man to my left and a white man, who was standing in front of a narrow table. The white man was the foreman and his name was Jim Davis. The black man was a worker, named Eddy.

"Hi . . . I'm Cher Waiters," I said with a little pep in my voice as I handed him my work order. "I was told to report here for work."

The room spun to silence as the men stared me down under the weight of a heavy scrutiny. Within moments, both men's eyes dropped to the floor and they both began to laugh aloud at the same time.

What? I thought to myself. *What's so funny?*

"Didn't they tell you what kind of boots to wear?" the foreman snapped. "You're supposed to be wearing Red Wings, *not* fashion boots." He lifted his foot high in the air so that I could see his left boot as he and the black guy shook their head and continued to laugh.

"Let's see what you have in your toolbox," he said turning his attention to my tools. By this time, he had made his way over to me, but the trailer was too narrow for both of us to stand in the same place at the same time; therefore, I had to back out of the trailer and stand on the steps while he reached his over-sized hands into my toolbox and began to pull out one tool after the other.

"You call these strippers?" he laughed, handing them to the Black guy, who seemed to be consistently amused by all that was going on.

"You call these side cutters?"

Eventually, he turned my toolbox upside down dumping all of the tools out—all the while shaking his head.

"Didn't they give you a tool list?" he barked.

"Yes . . . but it didn't specify what kind of tools to get," I offered in my defense.

"You'll need a pair of these," he said reaching in his back pocket and pulling out a pair of Klein side cutters.

"Okay," I said with a nod. I stood on those steps like a three year old child. I felt humiliated, raw and small. My skin felt as though the flesh had been picked off the bone—by hand.

"Let's go and do some work," the foreman insisted.

Blistering and outnumbered by the monstrosity of male ego, I allowed the men to get ahead of me and I turned to follow them. As we walked, the only comfort that I could find was in the fact that I was taller than the black man. I could feel devastation setting in and as we passed by a window, I happened to catch my reflection, and it was there that I saw my tormented reflection.

I looked horribly sad.

I wanted to cry.

Wither.

Sink.

No . . . just die and get it over with.

But I couldn't.

At least not here and not now.

Only a *girl* would do that—and if there was ever an hour whereupon I needed to be more than a girl, more than a woman, and definitely more than me—that hour had surely arrived.

And feeling every ounce of insecurity that I had ever had in my entire life as it rose to the surface, I quietly mumbled beneath my hard exterior walls, *"What have I done to myself?"*

But there was no one to answer me and the answers would not come for a long, long time. Therefore, in the interim—I followed quietly behind the men and died one thousand deaths while smiling.

When we reached the jobsite, I felt the door opening to a whole new world. It was a peek behind the curtain that women rarely, if ever, get the chance to behold. I felt like a member on the team of scientists in Jurassic Park when they first entered the grandiose world of dinosaurs.

I was amazed.

The structure was beautiful and even more impressive was the story "behind" the building. A minister had spent years saving the money that his congregants donated. With the funds he had acquired, he was building a nine million dollar apartment building, along with a conference hall for the members.

Impressive indeed, I thought to myself as I smiled.

Once we entered the building the first thing my eyes caught sight of was a tall, thin man who bent down, picked up a steel beam and carried it effortlessly away on his shoulders.

Wow!

Now that's strength!

He looked like something out of a superhero movie as he quickly faded out of my view. There were so many men with strong backs and bulging muscles, dripping in sweat, grime and testosterone. They were stretching, bending, reaching, pulling and carrying what seemed to be the weight of the world on their shoulders. There was only one black male worker on the whole job site Eddy and not a single woman or an ounce of femininity within one thousand miles of here. This could have been a woman's paradise, a cove of forbidden pleasures—if the ego-crazed men who comprised the entire population of this place hadn't been straight up assholes *(for lack of a better word.)*

The foreman led me to an area that would ultimately become the meter bank room. He left me with Eddy and a list of instructions for my co-worker. "Put in an eight foot trough, punch holes in it for the pipes coming out of the ground and for those going up to the meters."

Generally speaking, the foreman lays out the work and his workers carry out the plan. Some foremen work and others do not. And who you end up with is the luck of the draw. And with that—Jim the foremen disappeared. Eddy began to take measurements of the pipes to transpose them on to the trough; however, I couldn't help but notice that his measurements were off. Eddy can't add fractions.

"Hey . . ." I said offering help, "your measurements aren't gonna come out right."

"I've been doing this kind of work for ten years!" he immediately snapped. "You're just an apprentice, which means you're lower than whale shit . . . so just zip it shut, sit back and learn something!"

Well, excuse the hell out of me, I wanted to say. Instead, I opted to take two steps backwards and allow him the grace of "mucking" up the job. This wasn't all that I stood back and observed. During break, Eddy and I returned to the trailer. Upon entering, the foreman looked at me and said, "You bring coffee?"

"No," I quickly responded. "I don't drink coffee."

"Well . . ." he blurted, "you will if you're gonna be working 'round here."

Eddy snickered beneath his breath.

"Say . . ." the foreman said to Eddy pointing at my toolbox, "Pick up those tools!"

Eddy jumped at the command and began to pick up the tools without protest. I was shocked. The white Jim sat back in his chair and kicked his heels up on the desk. I observed in silence. It was almost short of unbelievable—the master/slave mentality. There was submission without questioning and a surrender of power without protest. The white foreman was rude and desperately lacked manners. The behavior was grotesque to watch and he seemed to be more cave man than anything else.

When break was over we returned to work. Eddy continued to muck up the measurements and I continued to stand by and watch. He was focused, determined and yet seemed desperately unhappy. In fact, all of the workers seemed unhappy.

Everyone worked.

Some even slaved . . . and everybody waited on their pay.

Union jobs paid very well, indeed.

Later that morning when Mother Nature called, I went in search of a restroom. Instead, I was directed to what is termed "the Redhead," (translation: portable potty). Needless to say, this portable restroom was disgusting and reeked from here to high heaven.

Pee You!!

"What in the world have I gotten myself into?" I asked again for the second time that day. By the time I returned to the area where I was working with Eddy, the foreman had returned and noticed Eddy's measurements were off. The foreman blew his stack and went to town in cussing Eddy out 69 ways to Sunday night.

"You stupid Son of a Bitch!" the foreman shouted, "Look what you did! Dumb ass!"

I couldn't believe what I was witnessing. Did the supervisors *really* speak to their subordinates in such a derogatory manner?

Oh my . . .

"What have I gotten myself into?" for the third time that day.

Everyone on this jobsite (including the foreman) could stand to retake a kindergarten class on etiquette. They walked harshly, talked harshly and breathed harshly. Under the weight of it all, I could barely breathe myself.

In a single day on the job, I observed racism.

Sexism.

Harassment.

And discrimination all rolled into one.

These were the rules of the jobsite, but the main rule was a simple one—*there are no rules.* Both the men and the job were dangerous. "Dumb ass Eddy," as he was affectionately coined by the Foreman, was laid off as a result of his incompetency.

Welcome to my world.

Again.

I had hoped that all I had witnessed upon this day wasn't "all in a day's work." But somehow, I began to get a queasy, nauseous feeling that somehow it was just *that.* However, despite all that I had witnessed, I wasn't as horrified as I should have been. Somehow, I believed that belonging to Local 38's International *Brotherhood* of Electrical Workers—would protect me somehow. I also presumed that there would be rules, guidelines and a protocol of etiquette to be extended to all who belonged. I was also under the assumption that they were working with me and for me . . . but never *against* me. In this way, I was pitted above the drama (at least I thought so). Imagine the rude awakening of a naïve, young woman who was destined to learn the ways of this "unique" subculture and the underground world of Labor Unions.

Every good story has an arch enemy or a damn good villain.

Welcome to the villain of my story.

Labor Unions have been defined as "legally recognized representatives of workers in an industry. Activity centers around collective bargaining over wages, benefits and working conditions for their membership."
Wikipedia encyclopedia

This sounds exquisite—at least in theory. However, my experience with the Union suggests that there is much more to the definition than could ever be recorded on paper.

They are a fraternity.

A brotherhood.

They think the same.

Act the same.

And are the same.

Filled to the brim with FBI *(translation: Fathers, Brothers and In-Laws.),* the Union is "organized labor" and operates in a very similar fashion to the Mafia. The Cleveland Union was a tricky one. They had a "separate but equal" philosophy. They segregated the minorities, which was reflected early on in my dealings with the Union—in that Black members had different training manuals than white members. Whites entered the Union as Apprentices. Blacks entered as "Trainees," which was the truest definition of "lower than whale shit."

The whole set-up was gangster from the very beginning. And even the "black teacher" (who had been hired by the Union to teach us) threatened to flunk me and four other minority students out of his class if we didn't pitch in and buy him a Rolex.

What in the hell?

In the face of such adversity, I organized a meeting at my house and rallied with fellow students to overcome such blatant and unacceptable treatment. On behalf of the minority students, I created the basis for an argument to plead our case. It was Supreme Court Justice Thurgood Marshall's "separate but equal" presentation, which we took straight to the business manager of the Union and got the teacher who tried to bribe us fired. But firing that guy didn't fix anything—it just ironed out one wrinkle in a crumpled suit.

Everything was slanted with the Union. Their practices may have appeared kosher to the untrained eye, but when scrutinized under a microscope—the story was quite different. Read between the lines: *unfair to the hilt.* When it was all said and done, the Union's "separate but equal" philosophy was an illusion. Blacks and especially women were separate—and *unequal* all the way down to our bone marrow. And I was destined to learn this firsthand.

Chapter 20

"I did what I did . . . to do what I had to do."

Cheryl Waiters

Day two on the job was better than day one. Primarily, because I had survived long enough, strong enough and good enough to show up again. This in and of itself was a magnificent accomplishment.

On day two, I arrived on time.

On day two, Jim the foreman was late.

On day two, I mustered up as much of a "good mood" as I could.

On day two, Jim the foreman seemed less than jolly.

On day two, I was still afraid.

On day two, Jim the foreman seemed fearless and I knew that he would cut a "man's balls off" if he so much as breathed out of the wrong side of his lungs—so I couldn't even imagine what he might do to me if I made even the slightest error in judgment. Therefore, I stood in rapt attention of every detail on the job—lest I miss something urgent and fall out of favor with those who ruled on decisions *and* destinies, both at the same time.

Upon arriving, I met the General Contractor and given my background and qualifications, he was quick to offer me a job with his company as a Project Manager.

"Thank you but no thank you," was my reply. I had my heart set on the Union's promise of $50,000 a year upon completion of my training and nothing was taking me off course from that promise. So, it was back to the business at hand—which I would soon dive into when Jim arrived.

"Come with me," said Jim as he took me inside of the building. As we walked, I observed Jim's behavior and I could tell he was a no nonsense kind-of-guy. He was a quiet man who did not come to the table boasting of many words. His language was limited, but his silent glare did much of the talking for him. Initially, Jim wasn't the easiest person to connect with, and if he did have redeemable qualities *(which I was certain he did)*—these were not advertised or put on public display for all to see. One would have to be a part of his interior world to get a better look at his heart.

"You ever used one of these before?" he asked, dragging out a power macho drill to the center of the room.

Woah!

My eyes nearly popped out of my head when I saw this massive, intimidating piece of machinery. This was the ultimate "boy toy" and it would almost seem to suggest that the operating manual for this power tool came with this general recommendation, *"for use by those who are pumped with testosterone only."*

"No," I responded to him honestly. "I've never worked with power tools before."

"Well . . . let me show you how to use this bad boy!" he offered.

I watched and listened intently. Jim wanted me to drill some holes that he had laid out and I had a pretty strong feeling that he was only going to show me how to do it *once.* His demonstration was quick and no nonsense. When he left, I knew that I was on my own and had to figure out a way to get the job done if I was going to make it back here and live to see day number three on this jobsite.

My first pass at the drill fell just short of disaster. I couldn't get a strong hold of it—the machine was just too powerful and heavy. I knew that I would have to find a way to handle the tool and its power. So, I stepped up my brain power and figured out a simple solution. I duct taped cardboard to my legs, which would enable me to have a buffer between myself and the machine. In this way, I could handle the tool's intensity without hurting myself. At times, I even had to sit on the drill—using all of my body weight to drill the holes into the floor.

I did what I did to do what I had to do.

I worked alone and no one was interested in being my co-worker *(much less my friend).*

I did what I was told and followed the law to the letter. I was meticulous, detail oriented and extremely organized. In many ways, I worked more efficiently and smarter than my male counterparts. I also had a secret weapon that most of my fellow workers did not have—I could read blueprints. This was crucial—even though only the foreman was allowed to access the prints, however, when Jim wasn't looking, I would sneak a peek and know what was just up the road ahead.

At the end of the day, Jim was impressed. When he wasn't focused on our differences, he warmed up to our similarities—and in doing so he realized that we shared common goals. These seemed to transcend gender differences. I, much like Jim, was interested in doing the best job I could and he respected me for that.

Ultimately, he took me under his wing and began to show me the "tricks of the trade." Jim gave me my first electrician's knife and even had heart enough to give me one of his jackets. He introduced me to the *essence* of electricity itself by allowing me to touch 110 volts of electricity with my bare hands just for the experience of knowing "what it felt like!"

Ouch!

Jim wanted me to know what raw electricity felt like so that if I came into contact with it on a ladder—that I wouldn't get too startled and fall off the ladder.

Yes, there was much to learn and sometimes the learning was both long and hard. And no matter how much Jim was taking a liking to me, I was still not welcomed on the site and the men resented me more with each passing day. When I interacted with the men individually, they were civil to me. When they grouped together and formed "herd consciousness" they behaved more like cattle *(i.e., animals)* and it was in this mindset that they treated me like shit.

I maintained my position by being precise in my work and my attitude. I was determined not to be broken—though they would bend me to the bitter edges on most days. My spine grew stronger and on the inside, I become like cast iron steel. I found a way to blend into the culture—at least with Jim. We found a way to balance "the yin and the yang." Jim would buy lunch and I would go and pick it up. Jim would buy bagels and I would toast them up. He even allowed me to bring in a nice blue tablecloth and some fresh flowers for that dusty old table in the trailer. Jim and I worked well as a team and I stayed with him for six months while he mentored me.

The experience was invaluable and when the superintendent, Marty, came around to re-visit our site, he was incredibly impressed with my progress. I had met Marty on my first day at the job and both he and Jim were probably a bit skeptical with regard to the nature of my longevity on the job. But I proved them wrong and by the time Marty came back around, he was amazed at how well I was able to work with the team. It was common knowledge on the jobsite that "Cher Waiters did the work of three men." And for that, she was despised by some and loved by others. In the long run, it served me well in keeping me in "high demand."

It was the spring of 1990 and I had finally completed my job with Jim. Shortly thereafter, I was transferred to The Kappa House, which was a 77-suite apartment building in Cleveland. I was working with an "admirable" guy named John. He was an engineer, could fly a plane and from what I could tell, he had "good home training." Good natured and gentle, I enjoyed working with John. However, the other guy *(a third year apprentice)* was the "mother of all mean co-workers." This guy was bad to the bone and a nightmare to deal with. He was out looking for trouble, living in trouble, starting trouble and being in trouble. He had a mean streak in him that was relentless. He liked to pull "tricks" on me. When I wasn't looking, he'd steal my tools and tear up my work. And on one occasion, while I was in the redhead (portable potty) the Bastard tried to tip the whole thing over! I went straight to John for help!

"John . . . you gotta help me with this guy! He's bad news!"

But no help would soon arrive. John did little more than appease me but nothing was ever done to get Hell Boy off my back. Again, I would have to be clever to continue on with my work so I devised a plan. I would work the apprentice to death. He was extremely competitive with me and couldn't stand to have a girl "show him up" on the job. If I wired two suits in one day, he would wire three. If I wired three suits, he would wire four. So, I started revving up my productivity and pretty soon, he didn't have time to play anymore tricks on me because he had to keep up and outdo me at every turn. I laughed to myself every day watching him race around like a "chicken without a head" trying desperately to keep up. It was hysterical and eventually before all was said and done, we had the suites all wired and the apprentice was transferred to another job.

Sweet victory!

Eventually, I, too, had to move on and was transferred to the Severance Mall Apartments. A guy named Dale was the foremen on this job and he was meaner than a "junkyard dog." From day one, the man did not like me and was on the verge of "ripping me a new one" till I was rescued by the long reach of my superintendent, Marty, who transferred me the hell out of there and back to John where the "living was easy" in comparison. As part of my training, I was assigned to a new contractor every six to eight months. However, a contractor could keep a Trainee (minority) for as long as he wanted to have him or her working for them. I was under the umbrella of *Hirsch Electric* through December of 1990 whereupon I was transferred to *Reserve Electrical Company* (Cleveland State University) in January of 1991. It was at Reserve Electrical that I became a real apprentice and was relieved of the bogus title "trainee." In fact, that was the end of the "Trainee Program."

We were the last of the Mohicans!

It was on this job at Reserve Electrical that I was befriended by an Italian foreman named Joe, who seemed to take a liking in me and respected my work ethic.

"What's a nice girl like you doing in my place like this?" he asked me.

Well, if that wasn't that the million dollar question at the end of the day, I didn't know what was!!!

"You want the long version or the short one?" I asked him.

"I only got three more years on this gig and I'm outta here . . . so you best give me the short version!" he said howling with laughter.

By the time I finished summing up the short version, he shook his head as he said to me, "it only gets worse, Cheryl."

"Yikes!" I shouted. "I thought things would get better."

"Lies!" he exclaimed, "All lies! This used to be a good trade but now it's gone to shit! The people . . . and the way they do things around here . . . back stabbers . . . cut throat . . . you'll be lucky to have a job. The only reason you're working now is because you're an apprentice . . . cheap labor."

I felt a lump forming in the base of my throat.

Nooooooooo! I wanted to scream.

Please don't burst this floating bubble!

I still got dreams of being somebody!

"What do you think happens when you top out in pay?" he further questioned me.

"I don't know . . ." was my only response.

"You'll be an overqualified apprentice and an inexperienced journeyman. Nobody will want to hire you . . . that minority female shit don't mean nothing here."

I was flabbergasted.

"You gotta make 'em think you're great," were his words of wisdom. "They think I'm great. That's how I've lasted this long."

"Well . . . I've been always been good at being great!" I reassured him.

I found comfort in greatness.

I always had and I always would.

"This used to be clean work, Cheryl . . . a respectable profession" he went on to tell me. "We wore white shirts and ties."

Wow!

I couldn't even imagine.

A white shirt and a tie was a world away from anything I had EVER seen in a day in the life of an electrician. The job itself was synonymous with dirt, grime and funk.

"We used to get a lot of respect," mumbled Joe, "Nobody respects us anymore. Look at these bums," he said pointing to a few slovenly dressed workers. "This ain't even a real job . . . it's the best part-time job in the world."

"What do you mean?" I asked him.

"Oh . . . you'll figure it . . ." he said, "but in the meantime . . . you got a long, long way to go."

The more he talked—the longer it felt.

Taking a breath and a pause, Joe pointed to a fifty foot extension ladder.

"Climb up there and take that pipe down," he instructed.

"Way up there!" I exclaimed under a bit of shock.

"Yeah . . . way up there."

Slowly, I began my ascent. There were no words to explain what I was feeling in climbing this ladder, but that's how I grew in confidence. Joe pushed me to be great and greater, a cut above the rest and a standout. As I climbed the ladder, Joe watched from the ground below.

"I gotta buff you up!" he barked. "Do some upper body exercises and put on some weight. You're too skinny!"

He talked.

I climbed.

He talked.

I climbed.

The more he talked, the higher I climbed.

Well, I thought to myself. *If I keep going . . . I just might make it to heaven.*

Chapter 21

"I left a trail and built a city!"

Cheryl Waiters

Summer of 1991

I continued to climb—not just tall ladders, but eventually, I made my way to the Society Building, which is Cleveland's tallest building in the city. This was a huge job and there were ten other guys working with me on the site. They were all white with the exception of one black man.

I was the only female on this gig.

Typical.

Oftentimes I was also the only black.

Also typical.

Interestingly, the white foreman on this particular job had an affinity for black women—which made for a "peculiar" work environment, especially when he gawked at me as though I were fresh meat on the platter of the week. I was the specialty item in all of the worker's eyes, but it didn't necessarily mean that everyone cared for the specialty. By the time I had arrived to the site, I thought I had pretty much seen it all *and* survived it all. *I've outlived the most brutal of all men,* I thought. However, when I was transferred to this job site, the statement "the cruelty of man" took on new life and meaning.

The guys resented me.

Now, I was used to resentment but their unfriendliness was layered with bitter sexism and oppression. None of the white workers on the site could stand the sight of me, so I was stuck with the black worker. Contrary to what one might think, there was no "unity" in the likeness of our skin. My black counterpart was a cowboy boot wearing, sex-crazed asshole *(for lack of a kinder adjective)*. Every morning upon my arrival at work, he would offer his own "unique" selection of sexual suggestive comments but not before sticking his tongue out and waving it at me.

Ughhhhhhhhhh.

Really?

The men on this site were one notch above Cannibals, if that. I had never seen such disdain, disrespect and absolute disgust of the female gender.

"Feed me, fuck me and leave me alone," one of the male workers boasted. "That's all women are worth."

Ughhhhhhhhhh.

Really?

"My old man told me to never trust anything that bleeds for seven days and doesn't die!" another co-worker blurted!

Ughhhhhhhhhh.

Again?

I hadn't seen anything like this. These men were brutal, raw and filthier than the dirtiest, ceiling I had ever seen. On one occasion, I reported to work and was asked to climb up into the ceiling to work on a box.

"You're small and light," I was told. "Climb that ladder and into that hole."

I wasn't crazy about the assignment, but nonetheless, I always did my job whether I liked it or not. So, I did as I was told. Upon completing my assignment, I tried to get out of the hole but the black worker had blocked me in.

"Move!" I demanded.

"Make me . . ." he suggested, trying to force himself upon me with a kiss.

"Move it!"

Again, he blocked my way.

He was serious. He wanted to play, but I was *not* a willing participant. He didn't care and I knew that he was the kind of guy who would take what he wanted. But what he didn't know was that I was serious too, and he wouldn't take a damn thing from me that I had not authorized to be released.

"Move!" I commanded one more time.

He didn't budge, so I kicked the ladder as hard as I could and the whole thing plummeted to the ground with him still on it. He hit the ground hard and cussed me 69 ways to Tuesday night when he got up off the floor.

"You Bitch!!!!!" he shouted.

"Your Mama!" I shouted back.

Tension was thick after that and I refused to work with him again.

"You can't refuse . . ." my foreman told me, "you're an apprentice."

"The hell I can't . . . I'll get my lawyer to solve this problem."

And with that—no one said another word. However, my black co-worker did make an attempt to get me laid off in retaliation for pushing him off the ladder. But at the end of the day there was no recourse—for him or for *me.*

During these days, I found strength, conviction and the courage to carry on in the legacy left behind by Harriet Tubman—one of the greatest abolitionist and humanitarians that the world had ever seen. I, too, had the strength and resolve of a woman such as Harriet Tubman. No matter what the situation or circumstance appeared to be, I refused to give up. Young and filled with energy, I pressed on . . . relentless *and* brilliant. The men I worked with were ignorant brutes. I had been trained by the military and had serious organizational skills. Filled with piss and vinegar, I did the work of three men while three men watched me "stuck on pause," because they couldn't believe what they saw in me.

I was a force to be reckoned with.

Always.

In my experience on the job and dealing with the union, situations of harassment, discrimination and sexism were not addressed and dealt with in the way one might think—especially when the situation involved a minority worker. Incidents that may sound "horrific" to the average person were swept under the rug and labeled *"all in a day's work."* No matter what went on during the course of a day—complaints were overlooked *(read between the lines: ignored)* lest someone actually had to do some work and resolve a filed grievance. And it was always a bit "sticky" when it came to the contractor and the workers. Ultimately, the contractor had to make certain that the job was done on time, within budget and done correctly. Therefore, more of their concerns were centered on keeping the male workers "appeased" so that they could finish the job. When blacks had disputes—the dispute was quietly dismissed. When the whites had a dispute—complaints were addressed.

Unfortunately, the ladder incident would not be the only one that I dealt with throughout my apprenticeship. I was sent from job to job and with each new jobsite, the situation grew in intensity and the men became more and more unbearable. The list of violations would include (but are not limited) to the following experiences, which I endured throughout my training and career as an electrician. The following were filed as grievances through the EEOC:

1. On the job, I was commonly referred to as "bitch" "cunt" or "douche bag."
2. I was consistently spoken to in a harsh, belittling manner.
3. I endured aggressive acts such as dumping my tool box contents on the floor, and stealing my tools.
4. A Journeyman spit on my car windshield, covering the window with mucous.
5. An apprentice endeavored to tip over the portable potty with me inside.
6. I was trapped in a ceiling by a fellow co-worker who attempted to fondle and kiss me.
7. Subjected to gross, foul and inappropriate sexual gestures and jokes.
8. A fellow co-worker asked me if I had ever been asked "to fuck from behind."
9. Consistent daily intimation and aggravation.
10. Damage to my vehicle (with intent to harm) when my truck tailpipe was bent under the back panel, which allowed exhaust fumes into the cab compartment.
11. A dummy torso of a woman with its head, legs and arms cut off was placed inside a Gang Box where my tools were.
12. A foreman sent me to work in a secluded area where no one else was around. He then proceeded to enter the secluded area to reveal his penis and make sexual advances toward me. I picked up a pipe with the intent to bust him in the head if he took another step toward me. *(Note: I was laid off by the foreman following this incident.)*

E.E.O.C. ignored the grievances *(translation: didn't do a damn thing about them).*

When I went to the Union seeking justice, I was told in no uncertain terms, "If the kitchen's too hot . . . get out!" Ultimately, upon review of the grievances by a federal judge—I was told only three offenses were serious enough to be taken further.

1. The attempt to tip over the portable potty with me inside;
2. Trapping me in the ceiling with the intention of kissing and fondling me;
3. The incident where the foreman attempted to reveal his penis to me in private and make advances.

The rest of the complaints were considered minor offenses.

When December of 1991 rolled around—the shit hit the fan. Somewhere between Thanksgiving and Christmas a heap of workers are always laid off, though there are a core group of workers who are always employed. Unfortunately, I was not a part of that group. In January of 1992, I was laid off and everything thing in my life came to a dead halt.

Everything.

Chapter 22

January 1992

"This ain't real and it definitely ain't what you think it is."

Cheryl Waiters

When I was laid off from Reserve Electric it seemed as though all hell broke loose and my eyes opened wide to the truth of how this thing, otherwise known as my professional life, really worked. Initially, the layoff was shocking and devastating. As I was still an Apprentice (translation: cheap labor), I was virtually guaranteed work, and rarely was an Apprentice laid off. But here I was—a jobless skilled laborer standing in the middle of what I referred to as "the next chapter of my life."

Note to someone somewhere . . . there's been a huge misunderstanding.

Nonetheless, I learned to stack up my pennies and add them up to dollars. And with that being said, somewhere between unemployment, sub-pay (a small monetary contribution to help one meet their expenses) and short call assignments, which were jobs less than two weeks in length, I managed to keep my head above water.

Barely.

In short, being in the Union *(which I was)* and working out of a Union Hall *(which I did)* was not the kind of gravy train ride that everyone around me assumed it was. This wasn't a corporate job; therefore "on-the-job security" was null and void. The system wasn't set up and designed to function that way. Electricians work when there's a project to be worked on.

When an electrician is in between jobs, he or she must go to the Union Hall where they sign in on the Union books. This ensures that they are added to the list of qualified professionals who are waiting for a gig. But an interesting thing about the Union is this—no matter what order you are on the list—if they don't like you . . . you don't get called. The Union is clever enough to devise creative ways to keep you unemployed.

Politics played a heavy role in dealing with the Union, and in my observation, minority workers were always the last to be hired and the first to be fired. And the only stability one could hope for once they made Journeyman was to become employed through a large corporation *(i.e., Ford, GM, the state government)* or any other entity that employed electricians full time on their roster. But everybody was trying to get those jobs and pickings were slim. When I got laid off it was a wake-up call, and no matter how grandiose the job sounded . . . it just wasn't real. Now, I was beginning to understand Joe the foreman's words, "this is the best part time job in the world!"

And with that knowledge—it was high time to get up, get out and mix it up with some miracles of my own. To that end, I had decided that in order to bridge the gap between jobs and to make my money "real" and "steady"—that I would purchase a family home, rent it out and become a landlord who would be on the receiving end of consistency.

Or so I thought.

I was in the midst of purchasing a three family home when I lost my job, but by the grace of a miracle and a little help from my friends—I was able to purchase a three family home in Cleveland proper. I was thrilled about my new purchase and the prospect of being a landlord with investment income. But before I could even boil my first cup of hot water on top of my new stove—the shit hit the fan.

Uh oh.

Neighbors from hell.

Interestingly enough, this dysfunctional group of people occupying the land next door to mine, were NOT a part of my closing escrow contract. The man, whom I purchased the house from—didn't mention a "peep" about what was really going on with the neighbors or with the neighborhood. And what was going on was the whole city of Cleveland was basically a ghetto and I had purchased a home on the outskirts of hell. There was a hotel at the end of the block (read between the lines: whorehouse) and a store with a big parking lot (translation: drug deals going down in the back.) Not to mention, my next door neighbor had 21 people living in the same house—with 16 of those people being the most disobedient, out of control, bad ass kids you ever wanted to see.

Holy Mother of Jesus!

The adults who lived in the house were not the most upstanding citizens. They made their living by pursuing the finer things of life (i.e. selling crack Cocaine, cussing and fussing, collecting welfare and intimidating the neighbors.) And from day one, we fought like the Hatfields & McCoys.

"Move out!" my neighbors shouted at me from across the fence. "You don't belong here! You belong in the country!!!"

Hell . . . maybe I did.

Their bad ass kids were like termites—tearing through my property and eating it up. They vandalized my property by spraying paint on it, putting their dog over into my yard and letting him dig it up. They would try to break into my house, steal my mail and throw bags and bags of garbage into my backyard.

Holy mother of Jesus!

With neighbors like this, what kind of tenants could I possibly get? That remained the most pressing question—which was most in need of an answer. My rental property was a revolving door and the tenants I acquired matched the décor of the neighborhood.

Shitty.

My life had become a living hell. I was battling with the Union in the daytime and fighting with those Mississippi Negroes that I called my "neighbors" at night. I had the local police department on speed dial, but unfortunately, when they did answer my calls—I was treated like I was the one on the wrong side of the law.

When I was growing up, my father had always told me, "Missy . . . there will be always be "Reggins" (the N-word spelled backwards) . . . stay away from those damn Reggins!"

But it was too late . . . I had already moved next door to them!!! And with each passing day, I fell deeper and deeper into despair. Still an Apprentice, I took what work I could get on short calls—but I wanted to make it right for myself. In the meantime, I had gotten wind of a new stadium known as Jacobs Field, a ballpark which would soon be under construction and the future home to the Cleveland Indians.

Yes!

Yes!

This is where I show them that I'm great and show them just what a woman can do!

There was something about the project that I found enchanting.

I'm going to make my mark with that project! I repeated over and over to myself. But in the meantime, I was doing the best I could on short term assignments. On one such assignment, I was working with a group of hard ass electricians . . . and one of the guys on the site continuously referred to me as "Bitch!" In fact, he never used my name at all. It was "bitch this" and "bitch that . . ." So, one day I had enough.

"Bitch!" he called out to me.

"Your Mama!" I retaliated.

What?

What?

The male co-workers were in an uproar over my response, and ultimately, they laid me off from this position. In that moment, it seemed as though it was just another story to chalk-up in the books—but it turned out to be a blessing in disguise.

After I was laid off, I returned to the Union Hall to sign in on the books again, letting them know that I was available for new work. When the Jacobs Field gig opened, I got a call and was sent to work.

Wow!

I would finally get the chance to show the whole world what I was made of, however, upon my arrival to the jobsite; I discovered that there was already trouble in paradise. The owners were threatening to throw everybody off the job because "Jacobs Field" was behind schedule. In addition, adjacent to Jacobs Field was the Gund Arena, which was the future home of basketball superstar, Lebron James. Ironically, the Gund Arena was also behind schedule.

My first day on the job, I hustled.

My second day on the job, I hustled.

I just kept hustling and allowed nothing to distract or deter me.

I felt as though I were building the whole thing by myself. The Superintendent was so impressed that he offered me a transfer to the Gund Arena—where I led a two man team and was instrumental in bringing in *both* projects on time! In fact, when I was working on the owner's suit in Jacob's Field, I had an opportunity to meet the owners personally and even put in a humble request. "Can I get two tickets for opening day of the Cleveland Indians?"

Well, no sooner than the words rolled off my tongue did the tickets fall into my hand. I was the proud owner of two tickets for opening day where president, Bill Clinton, was slated to throw out the first pitch of the game.

These were the glory days.

I had gone the distance. And given the fact that that I was the only woman who had worked on both Jacobs Field and the Gund Arena, I had the chance of a lifetime brewing beneath my Red Wing Shoes. And sooner than I had imagined—someone would come calling for me.

The only question was where and when? Well, the fantasy unraveled like this:

The City of Cleveland had promised to have a "minority quota" on the Stadium and Arena gigs. Therefore, in accordance with the law they were to employ 6.9% female workers and 33% minority workers. To make certain the city was in compliance, an organization known as the "Hard Hatted Women's Organization" was set in place to monitor the city's female participation in the project.

Shortly after my arrival, I was introduced to the organization's representative—a Caucasian woman who hailed from New York City with her own office on-site. She was a savvy businesswoman with a handful of "know how" and a reasonably "good head" placed upon her shoulders. Upon an introduction, we became "fast friends" in the sense that we were both friendly to one another. Ultimately, this woman not only befriended me—but she also believed in me. In fact, she supported me so much that she made it easy for me to believe in *her* when she came knocking in search of a $2,500 loan to help establish a Scholarship Fund for women in non-traditional fields.

I believed in the cause so I offered my support by writing a check—which came with three promises from the fast talking New Yorker:

1. To be remembered for my kindness;
2. To be repaid for my loan;
3. And to appear as a guest on the nationally syndicated show "Good Morning America" who was planning to feature a segment on Jacob's Field highlighting the women who had made contributions to both the stadium and the arena *(I was the only woman who satisfied this requirement.)*

At the end of the day—the New Yorker made good on her promise and I attained one of the greatest highlights of my professional career appearing on Good Morning America on April 4, 1994 in recognition of my journey as a female electrician in a male dominated trade, along with my contribution to Jacob's Field.

Camera.

Lights.

And action.

On Good Morning America and as I stood in Jacob's Field . . . I took my greatest bow.

The moment—

It was magical.

And the honor was more than I had imagined it to be.

I made my mark!!!

Following the completion of Jacob's Field, I became a Journeyman and topped out in pay. I did what they never thought I could do! And who are they? Well . . . just about *everybody.*

There truly was something "mighty" that lived within me.

I had poured out of my blood.

I had peeled back my skin and offered to the ground a hard days' sweat.

And now at long last, I was ready to put on some high heels and go on about the rest of my life as a woman.

I am strong.

I am invincible.

I am woman.

I built a city and a home for King Lebron James. I had walked a jagged road filled with adversity and had overcome a lifetime of pain and disappointment. Emerging from the womb during the "best and worst of times," I transcended the ignorance of a house filled with boys, a father who didn't know better and a mother who could never seem to do any better. I had fought my way beyond the borders of limited thinking on what a woman can and cannot do. I broke barriers, tore down walls and annihilated lifetimes of misunderstanding on the real value of a woman's worth. I did it with the grace of an eagle and the strength of a giant. I am a walking, breathing testimonial that a woman really *can* do anything that a man can do in her own special way, and in most cases, do it better. And ironically, sometimes the best "man" for the job truly is a woman—but it's not about hating on another or stirring up competition between the genders. It's about bringing the strength of each one to balance out the weakness of both. That is the truest definition of unity . . . at least in my book of *Blood, Sweat and High Heels.*

Indians

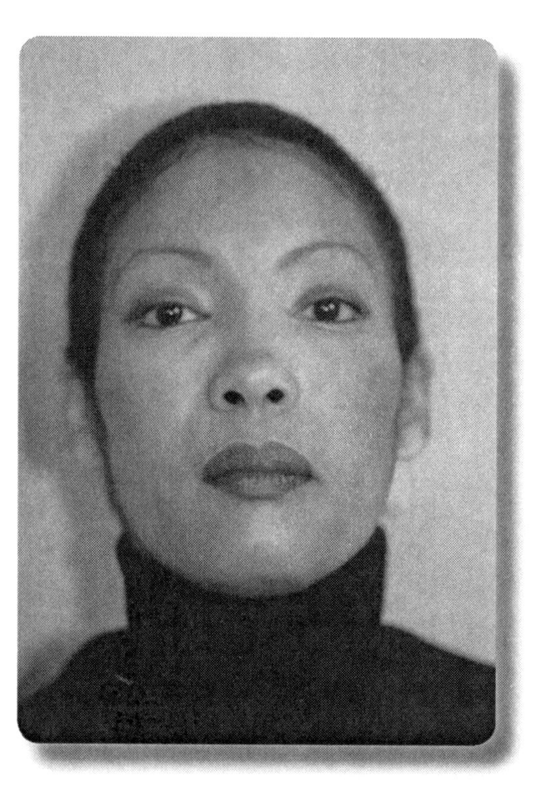

Epilogue

Cheryl Waiters has been a professional electrician for the past twenty two years. She topped out as a Journeyman and holds the distinction of working on some of America's most important landmarks. She currently resides in Cleveland, Ohio and continues her work as an electrician to this day.

You may contact Cher Waiters at cher4real@netzero.com

- African women architect
- womens history
- college - PU. stem careers, minority in their major, science
- women band lifted on active combat -> womens perspective/ empowerment / former military (general asso, womens interest
- sexual harrassment in workplace -> biz, women, counseling
- management -> women in workplace.
- military / career choosing patriotism over family
- Obama / black prez minorities 2nd term

CPSIA information can be obtained at www.ICGtesting.com
Printed in the USA
LVOW042248250113

317310LV00003B/96/P